Servicing Electronic Systems

Volume 2, Part 1

Basic Principles and Circuitry (Core Studies)

A Textbook for the
City and Guilds of London Institute Course No. 224 (as revised 1989–90)
and for
The Technician Education Council Level II Course in Electronics

by
Ian R. Sinclair
B.Sc.
Formerly Lecturer in Physics & Electronics
Braintree (Essex) College of Further Education

and

Geoffrey E. Lewis
B.A., M.Sc., M.R.T.S., M.I.E.E.I.E.
Formerly Senior Lecturer in Radio, Television and Electronics,
Canterbury College of Technology

AVEBURY

Second edition published 1991 by
Avebury,
The Academic Publishing Group,
Gower House,
Croft Road,
Aldershot,
Hants GU11 3HR,
England

Gower Publishing Company,
Old Post Road,
Brookfield,
Vermont 05036,
USA

Reprinted 1993, 1996

British Library Cataloguing in Publication Data

Sinclair, Ian R. (Ian Robertson)
    Servicing electronic systems
    Vol. 2. Pt. 1–2nd ed.
    1. Electronic equipment. Maintenance and repair
    I. Title   II. Lewis, Geoffrey E.   III. Sinclair, Ian R.
    (Ian Robertson) Electronics for the service
    engineer
    621.3810288

    ISBN 1 85628 167 1

Printed and bound in Great Britain by
Biddles Ltd, Guildford and King's Lynn

Contents

Preface

As the end of the century approaches, the technology of electronics that was born in the twentieth century is now the dominant technology in all aspects of our lives. The very nature of electronics has changed enormously in our life-times, from its beginnings in radio to its involvement in control of everything from food mixers to car engine performance, from games to industrial empires.

All of this makes the task of servicing electronic equipment more specialized, more demanding and more important. Servicing personnel play a very import-ant part in maintaining the correct operation of a system. They not only need to develop a high level of diagnostic skills but they also need to be able to communicate their findings to others so that the reliability and testability of a system can be improved. This then may also demand the further skills required to modify an in-service system. In particular, the training of anyone who will specialize in servicing must be geared to the speed and nature of the changes that are continually taking place. Such training must include a sound knowledge of principles and the development of diagnostic skills, neither of which is likely to be superseded by any changes in technology. Another important factor is that with the increasing harmonization of technical standards in Europe, it is likely that knowledge of technical terms in several European languages will become an essential part of the training for servicing work.

Although this series of books is designed primarily to cover the most recent requirements of the City and Guilds of London Institute Course No. 224 in Electronics Servicing, and also to provide coverage of the equivalent BTEC course, the books have been written mindful of the special needs of the home-based, distance learner. The approach is systems-based, viewing each electronic component or assembly as a device with known inputs and outputs. In this way, changes in technology do not require changes in the methods and principles of servicing, only to the everyday practical aspects which are continually changing in any case.

We have also taken every opportunity to look beyond the confines of the present syllabuses to the likely requirements of the future, and particularly to the impact of a single European market on both electronics and training. The books will invariably be amended in line with changes in the syllabuses and in the development of electronics, but the aim will be at all times to concentrate on the fundamentals of diagnosis and repair of whatever electronic equipment will require in years to come.

A guide book has been prepared which contains useful course hints and comments on the questions included in the main text. This booklet, which may be freely photocopied, is available free of charge to lecturers and instructors from:

Customer Services,
Avebury Technical,
Gower Publishing Co. Ltd,
Gower House,
Croft Road,
Aldershot,
Hampshire, GU11 3HR.

Other books in the *Servicing Electronic Systems* series that are in the course of preparation are:

Volume 2   Part 2 *Servicing Electronic Systems* (Television & Radio Reception Technology)
Volume 2   Part 3 *Servicing Electronic Systems* (Control System Technology)

1 Measurements and readings

Summary

Meter types. Effect of a meter on a circuit. Digital meter principles. Current and voltage readings. The CRO. Time and amplitude measurements. Tables and Graphs. Waveforms.

Since the servicing of all electronic circuits calls for the use of measuring instruments, it is necessary to be able to make effective use of these and of the readings obtained from them.

The two most important types of circuit measurements are d.c. voltage readings, using the multimeter, and signal waveform measurements, using the cathode-ray oscilloscope.

One preliminary point of importance is that, whatever the type of measurement made in an electrical or electronic circuit, the very act of connecting the measuring instrument into the circuit is likely to have some effect on the circuit, and so to affect the reading obtained.

Multimeters

Multimeters, as their name suggests, are instruments capable of measuring several ranges, usually of d.c. voltage and current, of resistance and of a.c. voltage. Only a few multimeters made nowadays include a.c. current scales, since readings of a.c. current can only be made if a current transformer is included in the meter.

1

(a) Analogue type

(b) Digital type

Figure 1.1   Meters

Both analogue and digital meters are in general use (see Figure 1.1). An analogue multimeter uses a dial and a pointer needle. The required reading is the figure on the dial directly under the tip of the pointer. Analogue multimeters use a moving-coil movement which itself takes some current from the circuit under test, to operate the meter movement.

Digital multimeters, on the other hand, contain no moving parts; the reading appears as a figure displayed on a readout similar to that of a calculator. A separate power supply, usually a battery, is used, so that practically no current is drawn from the circuit being tested. Either type of meter will have a measurable input impedance whose value will affect the readings.

Exercise 1.1

Connect the circuits shown in Figure 1.2 and use first an analogue and then a digital multimeter, each set for a 10V range, to measure the voltage V. Note the results, and check by calculation that the value of V ought to be 4.5V when the

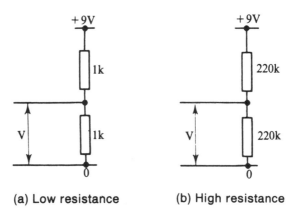

(a) Low resistance

(b) High resistance

Figure 1.2   Potential divider networks

meter is not connected. Are the meter readings significantly different for (a) the low-resistance circuit, (b) the high-resistance circuit?

The effect of meters on circuits

As the results obtained in Exercise 1.1 show, the lower resistance of the analogue meter can cause the readings it takes to be unreliable. These unreliable voltage readings are especially apt to be obtained when the resistance of the meter itself is not high compared with the resistance across which the meter is connected.

Figure 1.3 shows a bias circuit for a transistor. If the voltage at the base of the transistor is measured by connecting the multimeter between the base of the transistor and supply negative some current will flow through the resistance of the multimeter because this resistance is connected in parallel with the resistor R_2 in the circuit.

For reliable readings, the resistance of the multimeter must be high compared with R_2. At least ten times higher is the minimum – yet even with a meter resistance ten times the resistance it is connected across, some difference between the true voltage (when no meter is present) and the measured voltage must be expected.

Example: A transistor bias circuit having a 9V supply consists of a 68k and a 15k resistor arranged as in Figure 1.3.

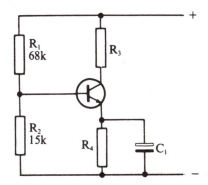

Figure 1.3   Transistor bias – circuit with potential divider

(*a*) What is the bias voltage on the base (assuming negligible base current)?
(*b*) What voltage will be measured by a voltmeter whose resistance is 150k?

Solution: The circuit is a potential divider.

(*a*) For such a circuit, $V = \dfrac{ER_2}{R_1 + R_2}$ with E = 9V, R_1 = 68k and R_2 = 15k.

3

$$\therefore V = \frac{9 \times 15}{68 + 15} = 1.63V, \text{ which is the base bias voltage.}$$

(b) The meter will be connected in parallel with the 15k resistor, so that the combined resistance will be $\frac{15 \times 150}{15 + 150} = 13.6k$. This quantity must now replace R_2 in the formula above, and the reading becomes:

$$V = \frac{ER_2}{R_1 + R_2} \text{ with } E = 9V, R_1 = 68k \text{ and } R_2 = 13.6k$$

$$\therefore V = \frac{9 \times 13.6}{68 + 13.6} = 1.5V.$$

The difference is 0.13V, which is about 8% low.

When the analogue meter is used for voltage readings, therefore, it is essential that the resistance of the meter for each voltage range be known.

The meter resistance can be found from the 'ohm-per-volt' figure, or 'figure-of-merit', which is printed either on the meter itself or in the instruction booklet covering its use. To find the resistance of the meter on a given voltage range, multiply the figure-of-merit by the voltage of the required range.

Example: What is the resistance of a 20kΩ/V voltmeter on its 10V range?
Solution: Meter resistance is 20k × 10 = 200k on the 10V range.

If the reading to be taken is across a high resistance, a high-resistance meter range must be used. Sometimes, it is even possible to use to advantage a higher meter range than the one which seems to be called for. For example, if a voltage of around 9V is to be measured and the 10V range of the meter has too low a resistance, the 50V or even the 100V range can be used to reduce the distorting effect of the meter on the circuit.

Note, however, that this would not be possible if the reading on the 100V range would thereby be made too low. A voltage of around 1.5V or less would be unreadable on the higher-range scales.

Digital multimeters generally have much less effect on the circuit being measured. Most of them have a constant input resistance of about 10M, and few circuits will be greatly affected by having such a meter connected into them.

Meters with still higher resistances are also available – up to several thousand megohms, if required. The operating principle is that the input voltage is applied to a high-resistance potential divider which feeds a *comparator* (see Figure 1.4). Another part of the circuit generates a sawtooth wave which is applied to the other input of the comparator. At the time that the sawtooth starts, a counter also starts and is stopped when the two voltages at the inputs of the comparator become equal. At that moment the count is displayed.

The circuit is arranged so that each digit of the count corresponds to a unit of voltage, say one millivolt, so that the display reading is of voltage. The range switch selects the part of the potential divider to be used, and the position of the

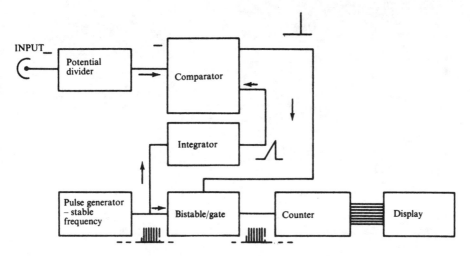

Figure 1.4   Block diagram of a digital meter (omitting details of hold-and-
measure cycle arrangements)

decimal point on the display. For example, suppose that the pulse generator or
clock of the digital meter runs at 1kHz, so that 1000 pulses per second are
generated. If the integrator is arranged so that each pulse produces a voltage rise
of 1mV, then the voltage will rise to 1V in the time of 1 second. For an input of
0.5V, 500 pulses will be needed, and the time needed will be 0.5s. The display
will then indicate a reading of 0.500V. Note, however, that the instrument has
taken 0.5 seconds to reach this reading. In general, digital meters are not
capable of following rapid changes in voltage.

Another point to note is that although a digital meter may indicate a voltage
reading to several places of decimals, this is not necessarily more precise than
the reading on an analogue meter.

Current readings

Exercise 1.2

Connect the meters, one digital and one analogue, to the centre-tap of the
potentiometer, Figure 1.5. Switch the meters to their 10V ranges and switch on
the 9V supply. Observe the readings as the potentiometer shaft is rotated to and
fro. Which meter most closely follows the changes in the output?

In most circuits, the use of a multimeter for current readings has much less
effect on the circuit than has its use for voltage readings, provided only that no
signal currents are present. To make a current reading, however, the circuit has
to be broken, and this is seldom easy on modern circuit boards.

5

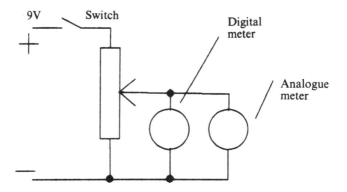

Figure 1.5   Comparing the response times of two meters

A few audio and TV circuits make provision for checking currents by having a low resistance connected to the current path. By measuring the voltage across this resistor, the current can be calculated using Ohm's Law:

$$I = V/R$$

Because the resistance so placed in the circuit is very small, the effect of connecting the meter into the circuit is negligible. Figure 1.6 illustrates these two ways of measuring current.

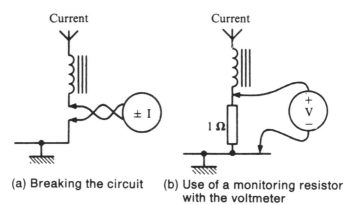

(a) Breaking the circuit   (b) Use of a monitoring resistor with the voltmeter

Figure 1.6   Current measurement techniques

Using the multimeter in circuits

The rules to be followed are these:

1 Start with the meter switched to its highest voltage range.
2 Connect the meter with the circuits switched off.
3 For voltage measurements, always use the highest range that gives a readable output.
4 For current measurements, always try the highest current range first.
5 Never leave the meter switched to any current range when you have temporarily stopped using it.
6 Make sure you know which scale on the dial to read before you try to take a reading.

The increasing popularity of digital multimeters is due to the clear readings of voltage which can be obtained by using them, combined with their high accuracy and the fact that they produce no reading problems and little or no disturbance of the circuit. Some of the more expensive of them are also self-ranging, meaning that no range switch is needed, simply a selector of voltage, current or resistance. The position of the decimal point is then controlled by the circuits inside the meter.

The a.c. voltage ranges of a multimeter are less often used, and should be reserved for checking mains voltage and the a.c. outputs from power supply transformers. These voltage ranges are scaled so that the meter reads r.m.s. voltage for a sine-wave input but will not (except for a few true-r.m.s. digital meters) read the r.m.s. value of any other waveform.

Because a moving-coil meter cannot read a.c. voltage (the average value of voltage over an a.c. cycle is always zero, which is what the meter would have to record), a *rectifier bridge* has to be used inside the meter for a.c. ranges. This rectifier, however, is inefficient both at low voltages and at high frequencies (certainly over 20kHz). The a.c. voltage ranges of the multimeter should therefore not be used as a substitute for the oscilloscope when measuring waveforms.

Voltage readings

Voltage readings are used to check the d.c. conditions in a circuit. These readings are therefore generally made when no signal is present. Readings taken when a pulse signal is present will obviously give misleading results because of the effect of the pulse signal itself on the meter. For this reason, voltage readings shown on a circuit diagram are usually specified as 'no-signal' voltage levels, or are shown only at points in the circuit where the signal is decoupled so that only d.c. is present.

In the linear amplifier circuit shown in Figure 1.7(a), the voltage reading at the emitter will not change greatly even if a small signal is present. In Figure 1.7(b), however, emitter voltage will be zero when no signal is present, so that the presence of any signal will cause a voltage of some sort to be recorded. Note

7

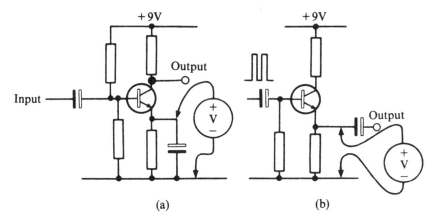

Figure 1.7   A linear amplifier circuit

that the voltage reading across a point which is bypassed by a large capacitor, as shown in Figure 1.7(a), will always be a purely d.c. reading.

When voltage readings are shown, they are always average readings, which may vary from one circuit to another of the same type because of component tolerances. Moreover, the actual readings taken in a circuit may be different from these values, either because of tolerances or because of the use of a meter with a different figure-of-merit, even if the circuit is working quite normally. Some experience is needed to decide if a voltage reading which is higher or lower than the stated average value represents a fault, or whether it is simply due to tolerances or meter resistance.

In general, a voltage varying from the norm by up to 10% can safely be attributed to tolerances; but a voltage variation that shows a transistor to be nearly bottomed or cut-off in a linear stage always betrays a fault condition.

Exercise 1.3

Use the circuit of Figure 1.8, in which the transistor is a general-purpose NPN type such as BC107/108/109, BFY50, 2N3019 etc. Connect the meter, digital or analogue as shown and switch to the 10V range. Switch on the d.c. supply, and read the meter. Now make readings with a squarewave input to the circuit from a signal generator, using 1kHz square waves starting at 0.5V p-p amplitude and increasing to 1.5V p-p amplitude. Observe and note the meter readings for different input amplitudes.

The CRO

The *cathode ray oscilloscope* (CRO) needs to be used in place of a meter of any kind in circuits (especially TV and industrial ones) in which voltage amplitudes,

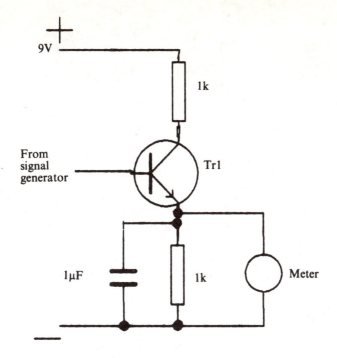

Figure 1.8   How signal inputs can change meter readings

wave-shapes and accurate pulse timing are more important than d.c. bias levels. In many such circuits, transistors are cut off or bottomed when no signal is present, so that bias readings are of little interest in any case, save possibly as a check.

Like the multimeter, the CRO will disturb the circuit into which it is connected for waveform voltage readings. Its input resistance is usually around 1M, unless a high-resistance probe is fitted. The trace is measured against the centimetre (cm) scale on the graticule – the lined transparent sheet (often coloured green) located over the screen of the tube face.

To measure the peak-to-peak amplitude of a signal, the vertical distance, in cm, between one peak and the opposite peak must be taken, using the graticule divisions. This distance (Y in Figure 1.9), is then multiplied by the figure of sensitivity set on the VOLTS/CM input sensitivity control.

To measure the duration of a cycle of a.c. from one peak to the next similar peak, the horizontal distance between similar peaks is taken, using the graticule scale and then the distance in cm is multiplied by the figure of time calibration on the TIME/CM switch. From the reading of wave duration for a complete cycle, the frequency of the wave can be found from the formula:

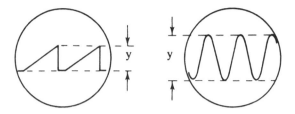

Figure 1.9   Measuring amplitude peak-to-peak

$$Frequency = \frac{1}{Time\,of\,one\,cycle}$$

With time measured in units of seconds, the calculated frequency will be given by the formula in hertz (Hz). If the time be measured in milliseconds (ms), the frequency will be in kilohertz (kHz); if in microseconds (μs), the frequency will be in Megahertz (MHz).

In some oscilloscopes, a continuously variable TIME/CM control needs to be turned to one end of its travel when time measurements are made, and any X-Gain control has to be at its minimum setting.

Exercise 1.4

Connect a signal-generator to the input terminals of the CRO. Set the signal generator so as to provide a 1kHz square wave of 1V p-p. Adjust the CRO so as to obtain a locked waveform, and read the amplitude and time. Does the time reading correspond to the frequency as set on the signal generator? Alter the signal generator amplitude and frequency settings, and measure the new waveform's amplitude and wave time on the CRO. Check that these settings match.

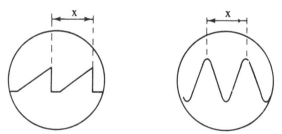

Figure 1.10   Measuring the duration of a cycle of a.c.

Using trigger inputs

An oscilloscope fitted with an *external trigger* (EXT.TRIG) input can be used for comparing the phases of two waves if a double-beam CRO is not available.

A wave into the EXT.TRIG input, with the trigger selector switch set to EXT., will cause the timebase to be triggered by that wave, so that the timebase always starts at the same point in the wave.

The triggering wave can be seen on the screen by connecting the Y-INPUT socket to the same source. The X and Y shifts can be used to locate one peak of the wave over the centre of the graticule (Figure 1.11(a)).

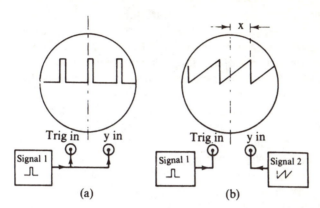

(a)   (b)

Figure 1.11   Using an external trigger to measure a time difference

If the Y-INPUT be now disconnected from the first wave and connected instead to a second source at the same frequency, a locked trace will appear on the screen (Figure 1.11(b)). If there is a time difference between the waves, the peak of the second wave will not be over the centre of the screen, for the timebase is still being triggered by the first wave. Measuring horizontally the distance X from the centre permits calculation of the time shift – either earlier (left of centre) or later (right of centre) – of the second wave as compared with the first.

This time difference can be converted into phase angle if the waves are sine waves. The conversion formula is:

$$\theta = \frac{360 \times t}{T}$$

where θ is the phase angle in degrees, t the time difference and T the time of a complete cycle expressed in the same units as t.

Example: A complete cycle of a waveform takes 3ms, and a second wave has its peak shifted by 0.5ms. What is the phase difference between the two waves?

Solution: $\theta = \dfrac{360 \times t}{T}$

11

$$= \frac{360 \times 0.5}{3}$$
$$= 60°$$

Capacitance and input resistance of a CRO

The input resistance of a CRO will disturb the d.c. conditions of a circuit if the circuit resistance is high (100k or so). The input capacitance of a CRO will also affect the signal waveform in the circuit if the circuit resistance is large.

The input capacitance of the CRO is generally around 30pF at the input socket, but will be much greater if a screened cable, or even a 'low-loss' coaxial cable, is used as a connector between the CRO and the circuit. An input capacitance of 100pF is quite normal when a fairly short length of such cable is used, and this input capacitance shunts the signal. If the signal comes from an output resistance of more than a few hundred ohms, the effect on the leading and trailing edges of square waves, or on the sine-wave frequency response, can be very noticeable. In effect, the combination of circuit output resistance and CRO input capacitance acts as an integrator or low-pass filter.

When the CRO is used to measure pulses in medium- to high-resistance circuits (a few kilohms or more), a low-capacitance probe should be used. Such probes are available as extras for most types of oscilloscopes. A few probes are active (i.e., they contain transistors or FET's), but most are passive, containing only a variable capacitor and a resistor (Figure 1.12(a)). The input voltage is

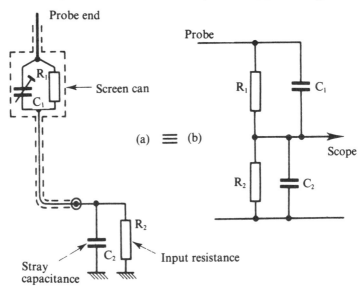

Figure 1.12   A low-capacitance probe

divided down, so that a more sensitive voltage range must be selected; but the effect of capacitance is greatly reduced because the capacitance of the cable and the CRO input is used as part of the divider chain.

In the equivalent circuit of a low-capacitance probe shown in Figure 1.12(b), C_2 includes the stray capacitances of the cable and of the oscilloscope. C_1 is varied until $R_1 C_1 = R_2 C_2$. Since signal attenuation is given by the expression $\dfrac{R_2}{R_1 + R_2}$, it is clear that a more sensitive oscilloscope range must be chosen.

Tables and graphs

Tables of values and graphs of variables are two ways of presenting information so that the eye can take in the information easily and quickly.

For comparing sets of voltage readings taken on a circuit, with one set normal and some sets indicating faults, the table is the best method of presentation. The voltage reading points should be arranged in order, and the readings shown clearly. The measuring units also given in Figure 1.13 show an example of such a table.

When a series of readings of one quantity is taken with its size varying with every reading, and it is desired to measure the effect that each change has on another quantity, the results can be shown in a table (or tabulated); but a graph is a much better method of showing the connection.

Say it is desired to measure the base current of a transistor over a range of values from 1μA to 100μA and to take a reading of collector current for each

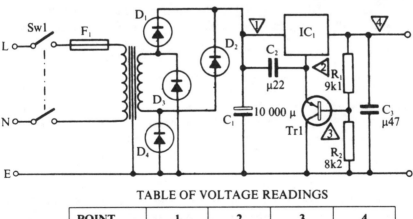

TABLE OF VOLTAGE READINGS

POINT	1	2	3	4
VOLTAGE	39	12	11.4	23

Figure 1.13   Voltage readings

value of base current. The results are first tabulated and then graphed. It will be found that the graph will show at once much more than can be gleaned after much effort from the table.

The tabulated readings shown in Figure 1.14, for example, do not show

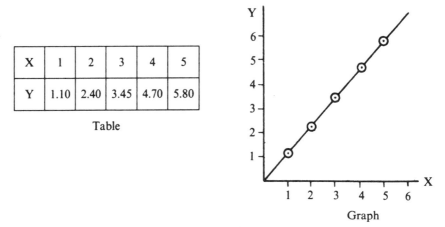

X	1	2	3	4	5
Y	1.10	2.40	3.45	4.70	5.80

Table

Graph

Figure 1.14   Table and graph readings

obviously what is immediately clear from a graph of the readings — which is that every unit change of the quantity X causes a unit increase of Y. This particular set of values gives a graph that is a straight line. There is then said to exist a *linear relationship* between the quantities X and Y. A resistor is an example of a linear device, and Ohm's Law is a way of expressing this linearity. This type of graph is the easiest to use, because the slope is constant.

The slope of a graph line is the ratio: $\dfrac{\text{Change of quantity Y}}{\text{Change of quantity X}}$ which, for a linear graph, has the same value at any part of the graph line. When the collector and base currents of a transistor, for example, are measured and a graph is plotted of collector current (on the Y axis) against base current (on the X axis), the resulting graph will be linear or almost so. The slope of this graph is the ratio h_{fe}, the *common-emitter current gain* of the transistor.

Exercise 1.5

Connect the circuit as shown in Figure 1.15. The transistor can be any general-purpose NPN type. Make certain that the potentiometer is set so that its voltage out will be zero when the circuit is switched on. Switch on, and take readings of Ib and Ic up to a maximum of 10mA collector current. Plot a graph of your readings, plotting I_c on the Y-axis and I_b on the X-axis. What type of graph shape do you think this is?

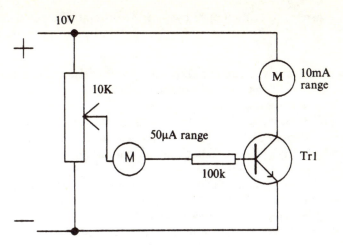

Figure 1.15   Circuit for measuring Ib and Ic for a transistor

Not all graphs, however, are linear in shape. The graph of voltage plotted against time for an a.c. wave, for example, is either a sine wave or a waveform of other shape. A graph of this type is repetitive, and is seldom plotted.

An important graph shape is the *exponential*. This is a graph whose slope values change in such a way that when a given amount is added to one quantity, the other quantity is multiplied by another factor. For example, the graph of the collector current, I_c of a transistor against the base-emitter voltage, V_{be}, is always exponential. For every 60mV increase (or decrease) in V_{be}, the current in the collector is multiplied (or divided) by a factor of 10.

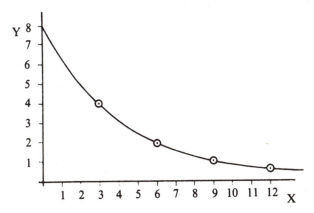

Figure 1.16   An exponential graph

It is typical of an exponential graph that the addition of a constant amount (in this case 60mV) to one quantity (base-emitter voltage) causes the other

15

quantity (current) to be multiplied by another constant factor (10).

Exercise 1.6

Connect the circuit as shown in Figure 1.17. The transistor can be any general-purpose NPN type. Make certain that the potentiometer is set so that its voltage out will be zero when the circuit is switched on. Switch on, and take readings of V_{be} and I_c from the value of V_{be} at which collector current just starts to flow, up to a collector current of 10mA. Plot a graph of your readings, plotting I_c on the Y-axis and V_{be} on the X-axis. What type of graph shape do you think this is?

An example of an exponential graph is plotted when a capacitor is charged or discharged through a resistor. It will be recalled that when a capacitor discharges through a resistor, the time constant RC gives the time for the capacitor voltage to reach 36.8% of its starting value. Because this is an exponential graph, when two time constants have elapsed (a time of 2RC) the voltage will be 36.8% of 36.8% of the starting voltage, or 13.5%. Similarly after the passage of three time constants, the voltage remaining across the capacitor will be only 4.9% of its starting value.

Here again, adding a quantity (the time constant) causes the voltage to be multiplied by another quantity (in this case 36.8%, or 0.368).

Another common shape of graph is the *inverse* graph, pictured in Figure 1.18. This type of graph is sharply curved in the middle of its range, but almost straight at each extreme, and the nearly-straight portions are almost parallel to the X and Y axes.

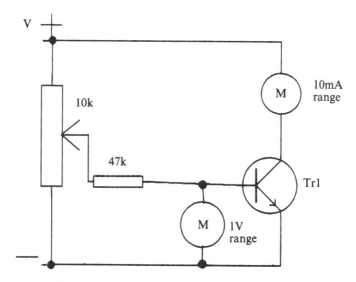

Figure 1.17   Circuit for measuring Vb and Ic values for a transistor

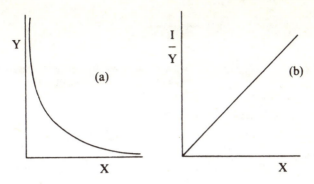

Figure 1.18   An inverse graph

An inverse graph results when one quantity depends on the inverse of the other. Take, for example, a simple d.c. circuit with a fixed voltage, and tabulate values of current I for different values of resistance R. A graph of I plotted against R will give an inverse graph, as shown in Figure 1.18(a).

If you recognize this as an inverse graph, you can check it by plotting another graph with one of the quantities (but only one of them) inverted. For example, plot I against $\frac{1}{R}$, or $\frac{1}{I}$ against R. The result (Figure 1.18(b)) would be a straight line.

Waveform graphs

A waveform graph shows the voltage (or current) that is present in the circuit *at each instant of time*. Because the quantity being plotted is a voltage (or current), there can be added to it any other voltage (or current) that is present at the same time and at the same place in the circuit. This is called *superposition*.

The simplest example of this occurs when both a steady voltage and a voltage wave are present in the same circuit. The value of steady voltage is added to the wave voltage at every instant, so shifting the whole graph upwards (if the steady voltage is positive) or downwards (if it is negative) by the amount of the steady voltage.

The voltages of other waveforms can also be added. The effect of adding a square wave to a sine wave is shown in Figure 1.19. Once again, voltages are added that are present in the circuit at each instant of time. Waveform addition of this type shows that combinations of sine waves, starting with a fundamental frequency f and adding components at the harmonic frequencies 2f, 3f, 4f, 5f, etc., yield other familiar wave-shapes such as *square waves* or sawtooth waves, all having the frequency of the fundamental wave.

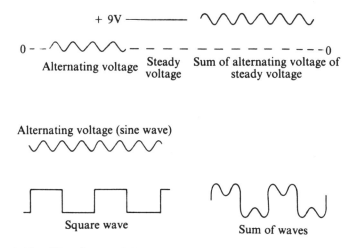

Figure 1.19   Waveform addition

Multiple-choice test questions

1   The use of a digital voltmeter is often preferable to the use of an analogue voltmeter because:
(a) it has an infinite value of input resistance
(b) it has a variable low input resistance
(c) it has a negligibly small input resistance
(d) it has a constant high input resistance.

2   In the circuit, Figure 1.20, the meter reads 9V on its 10V range because of a fault in a component:
(a) the resistor is short circuit
(b) the capacitor is short circuit
(c) the diode is open circuit
(d) the diode is short circuit.

3   In the circuit, Figure 1.21, the meter reads zero volts on its 10V range because of a component fault:
(a) the diode is open circuit
(b) the capacitor is short circuit
(c) the resistor is short circuit
(d) the diode is short circuit.

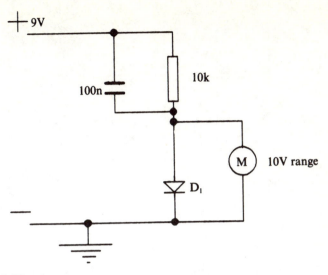

Figure 1.20

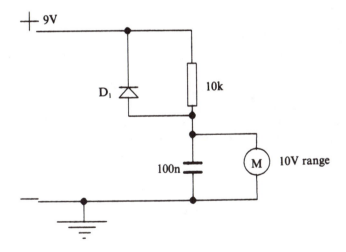

Figure 1.21

4 The graph, Figure 1.22, shows a waveform on the screen of a CRO which is set for 1V/cm and 1ms/cm. The sawtooth wave is of
(a) 2V p-p and 500Hz frequency
(b) 2V p-p and 2kHz frequency
(c) 1V p-p and 2kHz frequency
(d) 1V p-p and 500 Hz frequency.

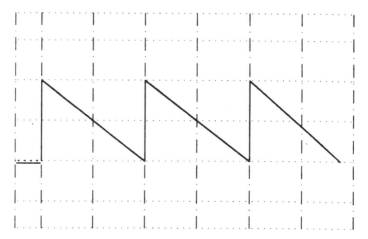

Figure 1.22

5 A graph shows that the current through a component doubles for each change of 0.2V across the component. This graph is
(a) inverse
(b) exponential
(c) linear
(d) sawtooth.

6 A fundamental frequency is
(a) the highest frequency detectable in a waveform
(b) the only frequency detectable in any waveform
(c) the middle frequency detectable in a waveform
(d) the lowest frequency detectable in a waveform.

2 Semiconductor devices (diodes)

Summary

Semiconductors and doping. Junctions. Diode Action. Diode characteristics. Zener diodes and LEDs. Rectifier circuits. The reservoir capacitor. Clipping, clamping and d.c. restoration. Amplitude demodulation.

Basic concepts

The two most important semiconductor materials, silicon and germanium, have resistivities whose values lie somewhere between that of a conductor and that of an insulator. The addition of very small amounts of 'impurities' such as the elements indium or phosphorus, however, can make both silicon and germanium conduct electric current. The process of adding impurities is called *doping*.

Heating a pure semiconductor also makes the material a better conductor, for as long as it remains at the higher temperature. Good conductors such as copper, and good insulators such as polythene, do not behave like this – neither doping nor heating having so much effect on their conductivity. The first and most important feature of a semiconductor material intended for use in electronics is that its conductivity can be accurately varied by the addition of small measured quantities of impurities.

Of even greater importance is the fact that not only can the conductivity of semiconductor materials be adjusted to any desired value, but that the type of conductivity can be pre-arranged. Most good conductors like copper and silver conduct because they contain a large number of particles called electrons which

21

are free to move between the atoms of the metal. These electrons are negatively charged, so that they always move from the negative pole of the battery towards the positive pole.

Another method of conducting electricity through solids has been known for most of this century – making use of the phenomenon known as 'holes'. A *hole* behaves in every way like a particle, but can best be described as 'the absence of an electron in a place where an electron ought to be'. In other words, the hole can be filled by an electron coming from somewhere else – this electron in turn leaving another hole behind it when it moves. Since electrons only move towards the positive pole of a battery, the sequence of holes created when electrons start to move in this way becomes a 'reverse flow' of positive current moving towards the negative pole of the battery.

Note carefully that holes exist only in materials of crystalline structure (like silicon and germanium); and that, unlike electrons, they never leave the crystalline material in which they belong in order to flow through a battery to begin their journey round the circuit again. A hole reaching the edge of a crystal is extinguished by being filled by an electron. To maintain the flow, a new hole has to be created elsewhere in the crystal itself.

Against this background – which is only a rough sketch of a complex physical process – can now be seen the two different types of impurity which are used to dope silicon or germanium. One type – it includes the elements phosphorus, arsenic and antimony – produces in the semiconductor material an excess of electrons which, when a battery is connected across the material, results in an electron flow (−) to (+) of the normal type. The electrons then become what is known as 'the majority carriers' – meaning that most of the current will be carried by them.

The other type of doping impurity – it includes the elements aluminium, gallium and indium – produces an excess of holes, which now become the majority carriers when the battery is connected and current flow begins.

A semiconductor material (say, silicon) in which the majority carriers are electrons is called *N-type silicon*. If indium, on the other hand, be used to dope silicon, the majority carriers will be holes and the doped material is called *P-type silicon*. The letters N and P are used because the electrons are negatively charged and the holes positively charged. Electric current flows in the material when these carriers move.

Junctions

N-type and P-type materials are not particularly important by themselves. But when a single crystal of silicon or germanium has both N and P-type doping, and the differently doped parts of the crystal meet, a *junction* is formed.

A junction of this sort is of great importance. A semiconductor junction that

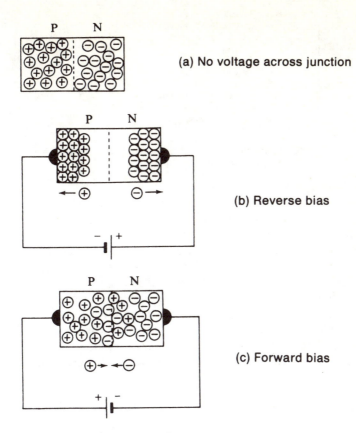

(a) No voltage across junction

(b) Reverse bias

(c) Forward bias

Figure 2.1   A semiconductor junction

forms part of a circuit will (at low voltages) conduct *in one direction only*. The action is illustrated in Figure 2.1, in which the + and − signs indicate positive and negative carriers respectively.

When the P-type material is connected to the negative pole of a battery, and the N-type material is connected to the positive pole of the same battery, both types of carriers are pulled away from the junction, which is said to be thus given *reverse bias*. With no charged carriers present at the junction, the junction becomes an insulator. The circuit is not complete and no current flows.

If the connections to the battery are reversed, however, and the junction is given *forward bias*, the effect is to pull both types of carrier across the junction so that the junction conducts and current flows in the circuit. This is how one form of semiconductor diode works.

Diodes can be made from any kind of semiconductor material. The most common materials used are germanium, silicon, and two-element materials such as gallium arsenide or indium phosphide.

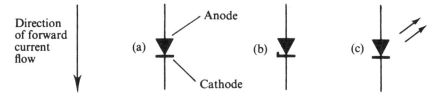

Note: diode symbols may be shown encircled

Figure 2.2   Diode symbols (a) Signal or rectifier diode (b) Zener diode (c) Light-emitting diode (LED)

Diodes may also have very small junctions – in which case they form what are called *point-contact* diodes. The larger junctions, of course, give rise to the name *junction diode*.

Diodes which are used only for their one-way conduction are called *signal diodes* or *rectifier diodes*. They are indicated by the symbol shown in Figure 2.2(a). Diodes used for voltage regulation, called *Zener diodes*, are indicated by the symbol in Figure 2.2(b); while diodes used to give a light signal, called *light-emitting diodes* or *LEDs*, use the symbol shown in Figure 2.2(c).

The arrowhead part of the diode symbol indicates connection to the part of the diode called the *anode*; the flat bar indicates connection to the part called the *cathode*. For normal (forward) conduction of current, the anode needs to be at a voltage more positive than the cathode.

Exercise 2.1

Use an ohmmeter to identify the anode and the cathode connections of an unmarked diode.

Note that, because of the internal wiring of the ohmmeter, the anode lead of the diode is that which is connected to the negative terminal of the ohmmeter when a low resistance is indicated. If a high resistance is indicated, reverse the connections to the diode.

A good diode should indicate a very high resistance when connected one way round, and a low resistance when connected the other way round.

Summary

Semiconductor materials such as silicon and germanium can be made into P-type or N-type conductors by adding small, carefully measured quantities of impurities to their crystalline structure.

Where P-type and N-type materials meet inside a crystal of semiconductor material, a junction is formed. Such a junction has diode action, in that it conducts in one direction only.

For such a diode to conduct, the connection that must be the more positive is the anode; the more negative connection is the cathode.

Signal and rectifier diodes

Signal diodes are used for demodulation, clamping and gating and are often point-contact diodes. These have very small junctions which can pass only small amounts of current; but they have the advantage that the capacitance between anode and cathode is also very small, which is a desirable characteristic of any component used in high-frequency circuits.

Junction diodes having larger-area junctions are better suited to rectifier circuits, which operate at low frequencies but pass large amounts of current.

Schottky diodes

A Schottky diode consists of a junction made between a pure metal, usually aluminium, and a N-type doped semiconductor material. Unlike a junction diode, this makes use of only one polarity of carrier (electrons). One very noticeable difference between a Schottky diode and a normal silicon junction diode is that the forward voltage for conduction is lower for the Schottky diode, given equivalent diode sizes. The other advantage is not so easy to understand unless we look closely at what happens in a junction diode when the bias is suddenly changed from forward to reverse.

In a P–N junction diode, during forward current there will be electrons moving through the P-doped areas and holes moving through the N-doped areas. When bias is suddenly reversed, these carriers are attracted back to their own regions, electrons to the N-doped region and holes to the P-doped region. This means that current will continue to flow for a short time (measured in nanoseconds) after changing to reverse bias. This continued current can be large, causing diode breakdown, or it can make the diode unacceptable as a rectifier for fast-changing signals.

Since a Schottky diode uses only one polarity of carrier, there are none of these 'stored charges' that need to return to specific regions, so that there is no continued current as the diode is switched suddenly from conducting to non-conducting. This makes Schottky diodes much more suitable for fast switching and for the demodulation of high frequencies or rectification of high frequencies.

Exercise 2.2

Connect the circuit illustrated in Figure 2.3, using a germanium diode.

Turn the potentiometer control so that the voltage across the diode when the

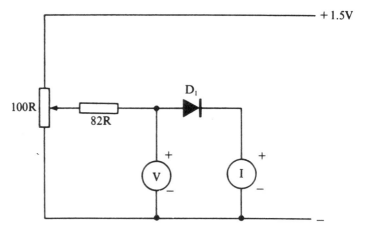

Figure 2.3   Circuit for measuring current through diode and voltage across diode

circuit is switched on is zero. Make sure that you know the scales of current and voltage you are using – a 1V (or 1.5V) scale of voltage and a 10mA scale of current. The voltmeter should be of the high-resistance type.

Watching the meter scales, slowly increase the voltage. Note the reading on the voltmeter when the first trace of current flow is detected. This value of voltage is called the *junction potential*. Below this value of voltage the diode does not conduct.

Note the voltage readings for currents of 1mA, 5mA and 10mA. Draw a graph and ask yourself if the diode obeys Ohm's Law ...?

Repeat the readings using a silicon diode. What differences do you notice?

Note, lastly, that the voltage across the ammeter (or current meter) can also be measured, but it will be very small compared to the voltage across the diode.

Characteristics of diodes

A resistor is specified by its values of resistance and power rating, so that asking for a 47k ¼W resistor ensures that the correct item is obtained. Diodes are less easy to specify because they do not obey Ohm's Law, and so have no consistently valid resistance value. To specify diodes completely, their characteristics have to be given – i.e., graphs of current plotted against voltage in both directions.

Typical characteristics for silicon and for germanium diodes are shown in Figure 2.4, illustrating the differences between these materials.

The characteristics show *voltage* plotted on the X-axis, with forward voltage (anode positive) plotted on the right-hand side and reverse voltage (anode negative) plotted on the left-hand side. Note that the scale used in plotting

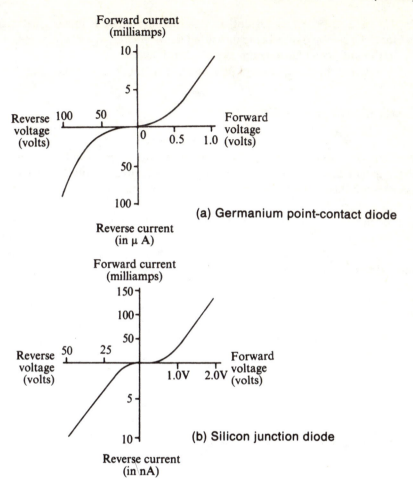

(a) Germanium point-contact diode

(b) Silicon junction diode

Figure 2.4   Diode characteristics

reverse voltage is very different from the scale used in plotting forward voltage because the voltage values differ so greatly.

Current in the forward direction is plotted on the Y-axis above the centre line, with reverse current plotted below the centre line, again on different scales because of the great difference in values. This method of plotting is always used for diode characteristics, so that the scale markings must always be examined very carefully.

The characteristics show clearly the differences between a germanium point-contact and a silicon junction diode:

1   A higher voltage is needed to make a silicon diode conduct. Typical threshold voltages are 0.5V for silicon, 0.15V for germanium.

2   The reverse (leakage) current of a germanium diode is much greater than
    that of a silicon diode, being measured in µA whereas the silicon is measured
    in the much smaller nanoampere (1nA = 1/1000µA).
3   A junction diode can pass higher currents than can a point-contact diode.

Zener diodes and LEDs

When a large reverse voltage (larger than the normal voltage across a conducting junction) is applied across a junction diode, the junction will break down so
that current flows. This breakdown occurs at a precise voltage whose value
depends on the amount of doping in the material and on the way in which the
junction is constructed.

It is therefore possible to manufacture diodes which will break down at fixed
and predictable voltages. Such diodes are called *Zener diodes*. They are used for
voltage stabilization purposes because the reverse voltage across a conducting
zener diode remains almost stable even when the current flow changes considerably.

Exercise 2.3

Connect a Zener diode in the circuit shown in Figure 2.5, making sure that the

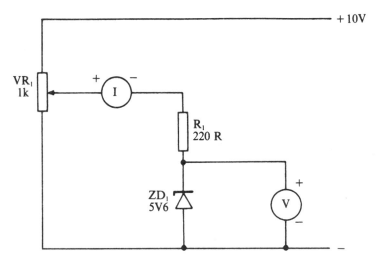

Figure 2.5   Zener diode measurements (Note the symbol for the Zener
diode)

cathode of the diode is connected towards the positive pole of the supply. The
voltmeter should be set to the 10V range and the milliammeter to the 10mA
range.

Switch on, and adjust the potentiometer so that the voltage across the diode can be read for currents of 1mA, 5mA and 10mA. Note the voltage readings for each of these current values.

If the first reading had been of voltage across an ordinary resistor, what would the voltage reading for 10mA have been?

(Hint – find the resistance values and then use the potential-divider formula)

LEDs (*light-emitting diodes*) are diodes made of transparent semiconductor material which makes visible to the user the light generated when current flows through the diode. The semiconductor material used has the property of emitting visible light when current flows through it.

LEDs are used as indicators of current flow. They need a higher forward voltage to make them conduct than do either silicon or germanium diodes. One notable feature of LEDs is that they have a very low reverse breakdown voltage, often as low as their forward voltage. LEDs must never be subjected to a.c. or to reversed-polarity d.c. of an amplitude exceeding their breakdown level. If in doubt, always connect a silicon diode in series with an LED.

Exercise 2.4

Connect an LED in the circuit shown in Figure 2.6. The voltmeter should be set

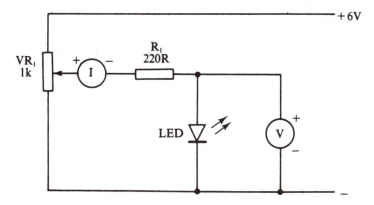

Figure 2.6   LED measurements

to the 5V range and the milliammeter to the 10mA range.

Starting with the potentiometer at its lowest voltage setting, switch on and slowly turn up the voltage until the diode conducts. Note the voltage across the diode which causes current to start flowing.

Examine the light output at current flows of 1mA, 5mA and 10mA, and note the values of forward voltage at these currents.

29

Selecting equivalents

When a diode has failed, either by becoming open circuit (o/c) or by starting to conduct in both directions (short-circuit, s/c), a replacement must be found. If the exact type required cannot be obtained, any equivalent must satisfy the following conditions:

1 The variety of diode (silicon or germanium, point-contact or junction) must be the same.
2 The maximum rated current of the replacement diode must be at least the same as that of the old one, or preferably somewhat greater.
3 The peak reverse voltage rating of the replacement must be the same, or greater.
4 The voltage across the diode at its normal operating current must be the same, or less.
5 Replacements for Zener diodes must have the same voltage and power ratings.
6 Replacements for LED's must have the same colour, forward voltage and current ratings as the ones they replace.

Summary

Diodes can be made of either silicon or germanium, and can be of either the junction or the point-contact type.

Germanium diodes have low forward voltage, but measurable leakage currents. Silicon diodes have higher forward voltages, but almost unmeasurably small leakage currents.

Zener diodes are given reverse bias, and begin to pass current only at their breakdown voltages, which can be predetermined.

LEDs are diodes giving a visible light output. They need a higher forward voltage than do germanium or silicon diodes.

Rectifier circuits

The circuit illustrated in Figure 2.7 is a *half-wave rectifier* circuit – so called because current flows through the diode only for that half of the a.c. wave which is in the forward direction. The output waveform therefore consists of the positive halves of the sine-wave only – with the result that the current output, although it is flowing in only one direction, is not smooth d.c.

A d.c. meter connected to the output of the rectifier will record a voltage of less than half the r.m.s. value of the a.c. voltage applied to the rectifier.

The pulses of current through a rectifier circuit are at line (i.e., mains supply)

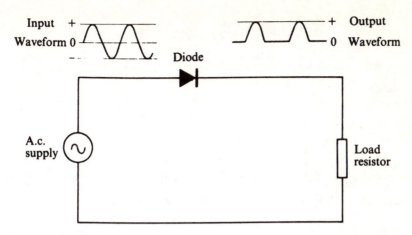

Figure 2.7   A half-wave rectifier circuit

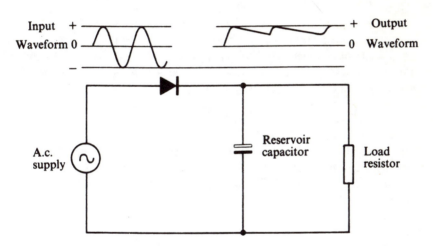

Figure 2.8   A reservoir capacitor added to the half-wave rectifier circuit

frequency, which is unusable for most electronic equipment. The waveform is greatly improved by inserting a *reservoir capacitor* into the circuit (Figure 2.8). As its name suggests, this capacitor stores some electric charge while the diode is conducting, then releases the charge to provide current flow when the rectifier is not conducting.

If a suitably large value of capacitance is used, current flow through the load resistor can be made continuous rather than coming in half-wave bursts. Moreover, this d.c. voltage will be much greater – almost equal to the peak voltage of the wave – if the capacitor be given a large value and its load current is small.

31

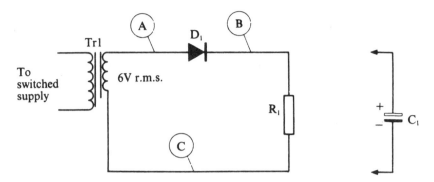

Figure 2.9   A measuring circuit for the half-wave rectifier

Exercise 2.5

Connect the circuit illustrated in Figure 2.9, in which D_1 should be an 1N4001, R_1 470R, ¼W and C_1 470μF, 25V. Set the oscilloscope to the 5V/cm input sensitivity range and the 10 ms/cm timebase speed, and switch on.

First, connect the oscilloscope earth to Point C in the rectifier circuit and its Y-input to Point A. Select a.c. INPUT on the oscilloscope. Adjust the controls to produce a visible trace, and switch on the a.c. to the rectifier circuit. If necessary, adjust the oscilloscope sync controls to produce a locked trace. Note the values of peak voltage, and the time interval between positive peaks.

Now disconnect the oscilloscope Y-input from Point A and connect it to Point B instead. Again note peak voltage and wave duration values. Observe the position of the flat 'base' of the waveform on the oscilloscope screen, and switch the input to d.c. By how many divisions on the graticule does the trace move upwards when this is done? What d.c. voltage does this indicate?

Connect a multimeter switched to the 10V voltage between B and C (B+ and C−). The reading shown will be of the average d.c. voltage at this point in the circuit.

Now switch off the rectifier circuit, connect in the reservoir capacitor C_1, and again take the readings between B and C.

Take great care that C_1 is correctly connected, with its positive terminal to the cathode of the rectifier. Capacitors of this type can explode violently if they are wrongly connected into a circuit.

The circuit shown in Figure 2.10 is that of a full-wave rectifier connected to a centre-tapped transformer. Because the centre-tap is earthed, the a.c. voltages at both ends of the secondary winding will be balanced about zero – i.e., when one voltage is at its positive peak, the other will be at its negative peak.

When the winding end marked X is positive, rectifier diode D_1 will conduct. When the winding end marked Y is positive, rectifier diode D_2 will conduct. In this way, both halves of the a.c. wave in turn are connected to the load.

The output voltage, measured with a d.c. meter, will be seen to be twice as high as the voltage output of the half-wave circuit (as would be expected). The frequency of the current pulses is twice supply frequency (100Hz instead of 50Hz, 120Hz instead of 60Hz).

Note that, for a given amount of load current, a smaller value of reservoir capacitor will be needed to smooth the output into d.c. of a voltage equal to the peak voltage of the wave.

The circuit illustrated in Figure 2.11 is that of a full-wave bridge rectifier, the most widely used type of rectifier circuit. No centre-tap is needed on the transformer because the arrangement of the diodes is such that the end of the winding which at any one moment is positive is always connected through the diodes to the same end of the load resistor.

Figure 2.12 shows the path of current through the diodes. In Figure 2.12(a), current from End X of the transformer winding flows during its positive half-cycle through D_1 and the load, returning to End Y during its negative half-cycle through D_4. When End X is on its negative half-cycle (Figure 2.12(b)), End Y is positive. Current therefore flows through D_2, the load and D_3, back to End X.

Thus the waveform across the load is that of a complete set of half-cycles in which both halves of the wave are used.

Exercise 2.6

Connect the bridge rectifier circuit shown in Figure 2.13, taking care that all the diodes – they should be 1N4001s or equivalent – are correctly connected. If one diode is incorrectly connected, it and the diode which is in series with it across the a.c. input will go open-circuit due to excess current.

Set the oscilloscope controls to 5V/cm and 10ms/cm, and switch the circuit on.

Examine the output waveform. Measure the d.c. voltage, using a multimeter on its 25V range connected across the load resistor. Remove any one diode, and study the effect this has on the waveform.

Now connect into the circuit the reservoir capacitor C_1. Again examine the

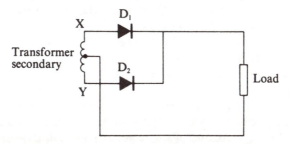

Figure 2.10   Full-wave rectifier circuit with centre-tapped transformer

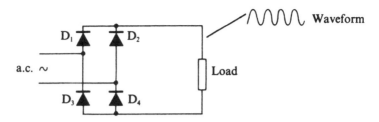

Figure 2.11   Full-wave bridge rectifier circuit

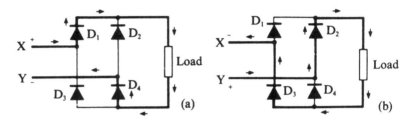

Figure 2.12   Current paths in the bridge rectifier circuit

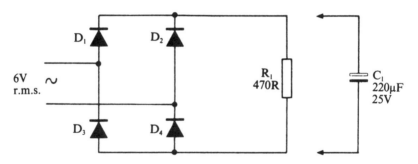

Figure 2.13   Bridge rectifier circuit

waveform across the load, and re-measure the d.c. voltage with the multimeter. Switch the oscilloscope input to a.c. and to a more sensitive range (say, to 0.5V/ cm) so that the remaining waveform (called the *ripple*) can be seen.

Finally, connect either a 100R, 1W resistor in place of the load resistor or a 220R, ¼W resistor in parallel with the existing one, and observe the change in the ripple waveform voltage.

Try the effect of removing any one diode from the circuit to simulate the effect of an open-circuit diode. Repeat your measurements of d.c. voltage, both with and without a load, and of ripple voltage on load. Explain why measurements taken on-load are more useful.

Rectifier circuits – operating characteristics

Circuit	D.c. output (No load)	Reverse voltage on diodes	Ratio I_{dc}/I_{ac}	D.c. output (Full load)
Half-wave	E	2E	0.43	0.32E
Full-wave	E	E	0.87	0.64E
Bridge	E	E	0.61	0.64E

The ratio I_{dc}/I_{ac} is the ratio of direct current flow through the load to a.c. input to the rectifier circuit.

In all rectifier circuits, each diode will be reverse-biased for half of the a.c. cycle and conducting for the other half. The amount of this reverse bias depends on the type of circuit used, and is greatest for a half-wave rectifier feeding into a reservoir capacitor.

The table that follows shows the operating results which may be expected from rectifiers of the three types described – half-wave, full-wave and bridge – given an a.c. input of E volts peak (0.7E volts r.m.s.) and smoothing by reservoir capacitor.

Note that in all rectifier circuits, reversing the connections of the diodes reverses also the polarity of the output voltage.

Summary

A half-wave rectifier uses a single diode, passing half of the a.c. wave. A full-wave or bridge rectifier passes two positive half-cycles in each cycle.

The addition of a reservoir capacitor brings the output voltage up to almost the peak a.c. value.

The ripple is at supply frequency when a half-wave circuit is used, and at double supply frequency when a full-wave or bridge circuit is used.

All diodes used must be correctly rated for forward current and reverse bias.

The Zener diode regulator

A Zener diode, as already noted, is used reverse-biased so that the junction breaks down to permit current flow through the diode. This breakdown occurs at a precisely calculated voltage depending on diode construction, and causes no damage provided the current flow is not excessive. To prevent this, current must be regulated by connecting a resistor in series with the diode.

The complete circuit is therefore that shown in Figure 2.14, with the output voltage across the diode being used to supply any other circuit or part of a circuit which requires a stable voltage. Many circuits, especially measuring and oscillator circuits, are adversely affected by voltage variation, which can be

35

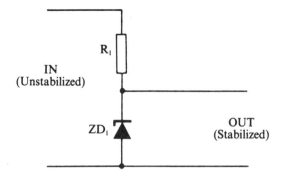

Figure 2.14   Zener diode regulator circuit

caused by changes in the supply voltage or by changes in the current drawn by the load.

Clipping, clamping and d.c. restoration circuits

All these signal-shaping operations (described in Volume 1) can be performed with the aid of diodes. A series-connected diode (i.e., one connected as for a half-wave rectifier) will pass only half of a waveform, so that the other half is *clipped* (Figure 2.15(a)).

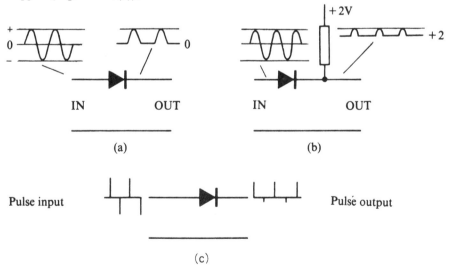

Figure 2.15   Clipping circuits

If a steady d.c. voltage is applied across the diode in addition to the a.c., a different part of the wave will be clipped. This makes it easy in practice to select any part of a waveform that may be required, and to clip the rest of it. In Figure 2.15(b), for instance, the addition of a resistor of the appropriate value allows selection of only that part of the waveform whose voltage exceeds 2V; while Figure 2.15(c) shows how (e.g.) the positive pulses only can be selected from a mixed $(+)$ and $(-)$ input.

Note that no capacitor is used in these clipping circuits. If the waveform needs to be fed through a capacitor, a resistor must also be added to prevent the capacitor from retaining charge after the diode ceases to conduct.

The operation of *clamping* restricts the voltage of a waveform to one set value for a short period. A simple diode clamp is shown in Figure 2.16.

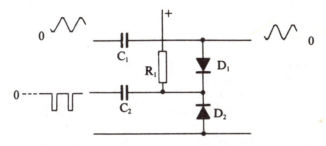

Figure 2.16   A clamping circuit

Because of the positive voltage at their cathodes, diodes D_1 and D_2 are normally not conducting. When a negative pulse is applied to the cathodes, D_1 conducts and the waveform at its anode is momentarily shorted through D_1 and D_2 to earth (ground). The inclusion of D_2 in the circuit ensures that the voltage at the cathodes of the two diodes is just below zero. If D_2 were not there, the clamping voltage would be the negative peak voltage of the clamping pulse, which might vary.

At the end of the clamping pulse, waveform voltage is free to vary again; but now any change of voltage will start from zero, because the capacitor C_1 has been charged through D_1 and D_2.

A *d.c. restoration* circuit is needed when it is necessary to hold some part of a waveform at a specified d.c. level. An example is the TV signal waveform, in which black level is represented by the cut-off voltage of the cathode ray tube. When the signal is either transmitted or passed through a transformer or a capacitor, the d.c. level is lost. An un-restored signal will then have a d.c. level of zero volts.

A d.c. restoration circuit can set the d.c. level of either peak of a waveform. Figure 2.17 shows one which restores the most negative part of a waveform to a level of 10V above zero. When any part of the waveform is at less than $+ 10V$, D_1 conducts and charges C_1 to a d.c. voltage of 10V. At one negative peak, C_1 will be charged so that the diode is only just conducting. With the negative peak

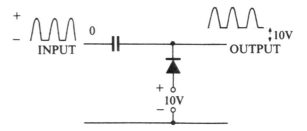

Figure 2.17   A d.c. restoration circuit

set to + 10V, the rest of the waveform will also be raised in voltage by the same amount.

Exercise 2.7

Construct the circuit shown in Figure 2.18. The diode is shown as type 2N4001, but any general-purpose silicon diode can be used. Set the potentiometer to give zero volts at the tap when the d.c. supply is switched on. Set the CRO on the 1V/cm scale (or nearest) and the signal generator for 1kHz square wave, 2V amplitude. Draw the waveshape as seen on the CRT screen, and then switch on the d.c. supply and adjust the potentiometer so as to alter the form of the waveshape. Draw the output waveshape for input d.c. voltage levels of 0.5V, 1.0V, 1.5V and 2.0V.

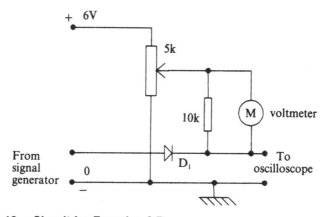

Figure 2.18   Circuit for Exercise 2.7

Diode protection circuits

Figure 2.19 shows three circuits which use diodes to protect other equipment. Figure 2.19(a) shows a circuit using germanium diodes connected across the

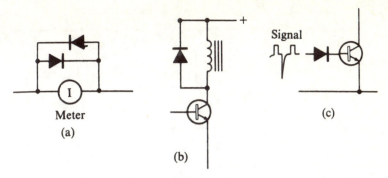

Figure 2.19   Diode protection circuits

meter movement to protect the meter against overloads. Each diode conducts when the voltage across it reaches about 0.15V, so that no more than this voltage can be applied across the movement in either direction. In most meters, the needle reaches full-scale deflection when the voltage across the coil is 0.1V or less, so that the diodes do not interfere with the normal action of the meter.

Figure 2.19(b) shows the circuit of a diode used to protect a transistor which is switching current into an inductive load such as a relay. When the current across such a load is suddenly switched off, the voltage across the load rises to a very high value (many times the supply voltage), which can damage the transistor. A diode connected as shown will conduct when the voltage surge occurs, so protecting the transistor.

Figure 2.19(c) shows a diode connected so as to protect the base-emitter junction of a transistor from reverse voltage. Many silicon transistors have a base-emitter junction which will break down on the arrival of a quite low negative pulse of voltage (10V or so). The connection into the circuit of a diode having a higher reverse voltage rating than this averts the danger.

Amplitude demodulation

The circuit of a simple demodulator for amplitude-modulated signals is shown in Figure 2.20. The action of the diode is similar to that of a half-wave rectifier, passing only half of the modulated radio signal wave. The output of the diode is a signal, still at radio frequency, whose amplitude varies. The average voltage at the output therefore also varies, at the frequency of the modulation.

The value of C_1 is small enough to present a high impedance to audio frequencies, but a low impedance to the radio frequencies. The load resistor across C_2 will allow C_2, which is charged by the rectifying effect of the diode on each r.f. peak, to discharge fast enough to follow the contour of the a.f. modulation. Examine the effect of removing the load resistor, and then try the effect of replacing the load resistor but removing C_2.

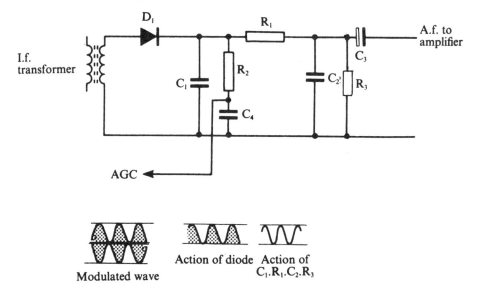

Figure 2.20   The diode demodulator

Since the diode has the normal action of a rectifier, d.c. is also present – with a voltage equal to the peak voltage of each r.f. wave if it be well smoothed. This d.c. is separated from the audio frequencies at the cathode of the diode by another low-pass filter (R_2C_4), and is then used to provide automatic gain control (AGC).

The capacitor C_3 prevents the d.c. voltage from affecting the bias of the next amplifying stage.

Exercise 2.8

Connect the demodulator circuit shown in the circuit of Figure 2.21. D_1 should be a germanium diode such as OA90, OA91, IN541, IN490 or IN127. Then connect the oscilloscope with its earth lead to Point D and its Y-input to Point A. Apply to the anode of the diode a signal at 1MHz, modulated at 400Hz. Set the Y-input sensitivity of the oscilloscope to 1V/cm and the timebase to 1ms/cm, and switch on. Adjust for a locked trace and examine the waveform.

Now clip the Y-input of the oscilloscope to Point B, and examine the new waveform which appears. Sketch the waveforms at A and at B, and then connect the Y-input to Point C, sketching this waveform also.

Now use a high-resistance multimeter set to the 2.5V range to measure the d.c. voltage between E and D. Compare the effects on this voltage of (a) varying the amplitude of the 1MHz signal only, (b) varying the amplitude of the 400Hz signal only.

If the 400Hz waveform cannot be varied, switch the modulation on and off alternately, and note the effect on the d.c. level.

The results will show, first, that the demodulated signal is the 400Hz modulating wave, free of radio frequency; and, second, that the AGC voltage depends on the amplitude of the r.f. carrier, not on the amplitude of modulation, when a normal a.m. signal is applied.

Summary

Diodes are used in voltage regulation (Zener diodes), voltage and current indication (LEDs), clipping, gating, clamping and a.m. demodulation. For the last four types of operations, signal diodes with small values of capacitance are preferred.

A diode may fail in either of two ways – *open circuit (o/c)*, when no current can pass in either direction, or (less commonly) *short circuit (s/c)*, when current passes freely in both directions.

Open circuit failures have the following effects:

1   The d.c. output of rectifier circuits is either reduced or falls to zero. A half-wave circuit will have no output; full-wave and bridge circuits will have reduced output, with ripple at supply frequency if only one diode fails.
2   A Zener diode regulator will give a higher voltage output, without regulation.
3   Clipping, clamping and gating circuits will not work. When the diode is in series with the wave, there will be no output signal. When the diode is in parallel with the wave, the signal will be unaffected by the diode.
4   There will be no signal from an amplitude demodulator, and no AGC voltage. No detectable r.f. will appear at the cathode of the diode.

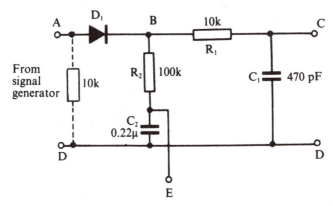

Note: In this exercise, shunting AD with a 10k Ω load resistor often produces better results.

Figure 2.21   Circuit for Exercise 2.8

Short-circuit failures of diodes have the following effects:

1 Rectifier circuits will blow fuses, and electrolytic capacitors may be damaged. A short-circuit diode will usually fuse itself, so becoming open-circuit. In a bridge circuit, a short circuit diode will usually cause another diode to become open-circuit.
2 A Zener diode regulator will give zero output.
3 Clipping, clamping and gating circuits will not work. When the diode is in series with the wave, the output signal will be identical to the input signal. When the diode is in parallel, there will be no output signal at all.
4 There will be no audio output from an AM demodulator, and no AGC. The still-modulated r.f. wave will then be present at the cathode of the diode.

Multiple-choice test questions

1 A piece of N-type silicon contains:
   (a) electrons only
   (b) equal numbers of electrons and holes
   (c) some surplus mobile electrons
   (d) some surplus mobile holes.

2 Which of the graphs of Figure 2.22 most closely represents the characteristics of a 1A silicon diode?

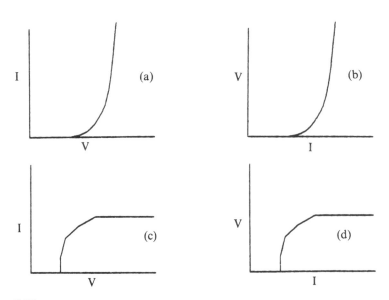

Figure 2.22

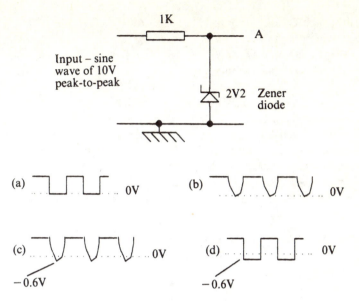

Figure 2.23

3 Referring to Figure 2.23, the waveform at the output of the circuit is most closely represented by the sketch in:

4 Figure 2.24 shows a diode rectifier circuit. When point A is positive and point B is negative, current flows:

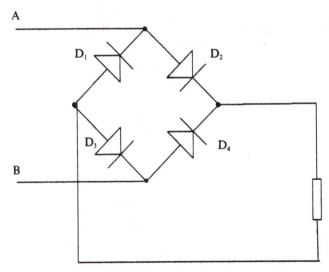

Figure 2.24

    (a) through D_1 and D_4 only
    (b) through D_2 and D_3 only
    (c) through D_1 and D_2 only
    (d) through D_3 and D_4 only.

5 Connecting a reservoir capacitor to a rectifier circuit will
    (a) increase the d.c. output level and increase the ripple
    (b) reduce the d.c. output level and reduce the ripple
    (c) reduce the d.c. output level and increase the ripple
    (d) increase the d.c. output level and reduce the ripple.

6 Referring to the circuit of Figure 2.25, the input is r.f. modulated by an a.f.

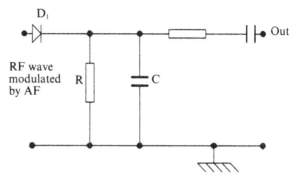

Figure 2.25

wave. The combination of R and C has a time constant which
    (a) must be large compared to the time of both the a.f. and the r.f. wave
    (b) must be large compared to the time of the r.f. wave but small compared to the time of the a.f. wave
    (c) must be small compared to the time of both the r.f. wave and the a.f. wave
    (d) must be small compared with the time of the r.f. wave but large compared with the time of the a.f. wave.

3 Transistors and other semiconductor devices

Summary

Bipolar transistors. Identifying connections. Current gain, input and output characteristics, mutual conductance. Transistor substitution. Failure of transistors. Field effect transistors. Junction and MOS types. Unijunctions. Thyristors, diacs and triacs. Hall-effect devices. ICs – construction, packages, removing and replacing.

Bipolar transistors

Bipolar transistors each have two junctions and three separate connections, as shown in Figure 3.1. The NPN transistor (a) has a thin layer of P-type material sandwiched between thicker N-type layers; the PNP transistor (b) has a thin layer of N-type material sandwiched between thicker P-type layers.

The layer which forms the middle of the sandwich is called the *base*; the other two are called the *emitter* and the *collector* respectively.

Figures 3.1(c) and 3.1(d) illustrate schematically the NPN and the PNP transistor, respectively. In Figure 3.1(e) the three connections of a transistor are named and indicated. The direction of the arrow-head on the emitter symbol distinguishes the transistor illustrated as being the NPN type. (The arrowhead points in the conventional direction of current flow.)

If the two junctions in any bipolar transistor were far apart ('far' in this case meaning more than a fraction of a millimetre), current flowing across one junction would have no effect on the other junction. Bipolar transistors are so made, however, that the junctions are very close to one another, so that electrons or holes moving across one junction will nearly all move across the other junction also (Figure 3.2). The result is that current flowing in one junction controls the amount of current flowing in the other junction. (Bipolar transistors, of course, are so called because both holes and electrons play their parts in the flow of current.)

As noted in Chapter 2, a junction is said to be forward-biased when the P-type material is connected to the positive pole of a battery and the N-type material is connected to the negative pole of the same battery of supply.

Imagine now an NPN transistor connected as in Figure 3.3(a). With no bias voltage, or with reverse bias, between the base and the emitter connections, there are no carriers in the base-emitter junction, the voltage between the collector and the base makes this junction reverse-biased, so that no current can flow in this junction either. The transistor behaves as if it were two diodes connected anode-to-anode (Figure 3.3(b)). No current could flow in the circuit even if the battery connections were to be reversed.

When the base-emitter junction is forward-biased, however, electrons will move across this junction. Because the collector-base junction is physically so close, most of the electrons will move across this junction also, so making it conduct even though it is reverse-biased. Any electrons passing across the base-collector junction are then swept towards the positive pole of the supply.

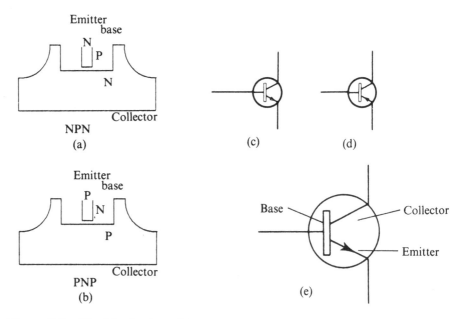

Figure 3.1   The bipolar transistor

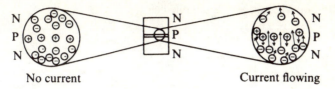

No current   Current flowing

Figure 3.2   Current flow in bipolar transistor

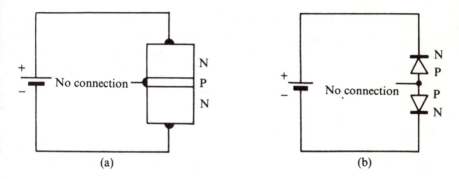

(a)   (b)

Figure 3.3   An NPN transistor with no base bias
(a) Schematic (b) Equivalent circuit

With both junctions conducting, most of the current will flow between the collector and the emitter, since this is the path of lower resistance. The transistor now no longer behaves like two anode-to-anode diodes because the electrons passing through the base-emitter junction make the collector-base junction conduct despite the reverse bias between collector and base.

The current flowing between the collector and the emitter is much greater (typically 25 to 800 times greater) than the current flowing between the base and the emitter.

If the base be now unbiased or reverse-biased again, no current can flow between the collector and the emitter. Thus the current in the base-emitter junction controls the amount of current passing through the collector-base junction.

Exercise 3.1

Using a silicon NPN transistor such as either 2N3053, 2N1711, 2N2219 or BFY50, measure the resistance between leads, using a multimeter set on the ohms scale. Remember that the multimeter polarity (+ and −) markings are reversed when the ohms scale is used, so that the terminal marked (+) is negative for ohms readings and the terminal marked (−) is positive.

Set out your readings as in the table below.

R_{bc}		R_{bc}		R_{cc}	
b+	b−	b+	b−	c+	c−

Note that each reading is taken in both directions so that, for example, R_{be} (b+) means a connection between base and emitter with the base (+), and R_{be}(b−) means the same connections but with the polarity reversed.

Note that if the connections to the transistor are unknown, the base lead can be identified since only the base will conduct to the other two electrodes with the same polarity.

When an NPN transistor is tested, the base will conduct to either emitter or collector when the base is positive. The base of the PNP transistor will conduct to both emitter and collector when the base is made negative.

Repeat the tests, and fill in a new table using a PNP transistor such as the 2N2905 or BFX39.

Then try to identify the leads of an unmarked transistor.

Tests with an ohmmeter can identify junction faults. A good transistor should have a very high resistance reading between collector and emitter with either polarity of connection. Measurements between the base and either of the other two electrodes should show one conducting direction and one non-conducting direction. Any variation from this pattern indicates a faulty transistor with either an open circuit junction (no conduction in either direction) or excessive leakage (conduction in both directions).

Current gain

It has been seen that the amount of current flowing between the collector and the emitter of a bipolar transistor is much greater than the amount of current flowing between the base and the emitter, but that the collector current is controlled by the base current. The ratio: $\dfrac{\text{Collector current}}{\text{Base current}}$ is in fact constant (given a constant collector-to-emitter voltage), and is commonly called the *current gain* for the transistor (its full name is the *common-emitter current gain*). The symbol used to indicate it is h_{fe}. A low-gain transistor might have a value of h_{fe} of around 20 to 50, a high gain transistor one of 300 to 800 or even more.

Note that the tolerance of values of h_{fe} is very large, so that transistors of the same type – even transistors coming from the same batch – may have widely different h_{fe} values.

Exercise 3.2

Measure the h_{fe} values for a number of transistors, using a transistor tester.
If a tester is not available, use the circuit of Figure 3.4. which will give

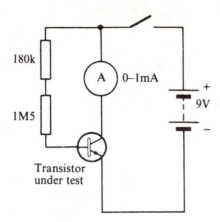

Figure 3.4   Circuit to test transistor

approximate h_{fe} values for a silicon NPN transistor by the current readings on the multimeter when put through the conversion table below:

Meter reading	h_{fe}	Meter reading	h_{fe}
1 mA	200	0.5 mA	100
0.9 mA	180	0.4 mA	80
0.8 mA	160	0.3 mA	60
0.7 mA	150	0.2 mA	40
0.6 mA	120	0.1 mA	20

These results are achieved because the two base resistors maintain current flow at about 5μA.

The characteristics of a typical silicon transistor are shown in Figure 3.5. These are the graphs which show the behaviour of the transistor.

Figure 3.5(a) shows the *input characteristic*, or the I_{be}/V_{be} graph. The slope of the line on this graph gives the inverse of the input resistance of the transistor, and its steepness shows that the input resistance is small. The fact that the graph line is curved shows that input resistance varies according to the amount of current flowing, and is greatest when the current flow is small.

Figure 3.5(b) shows the I_{ce}/I_{be} characteristic, called the *transfer characteristic*. This graph is a nearly straight line whose slope is equal to the current gain, h_{fe}.

Figure 3.5(c) shows the *output characteristic*, I_{ce}/V_{ce}, whose slope gives the value of output resistance. The horizontal parts of the graph lines show that a change in

49

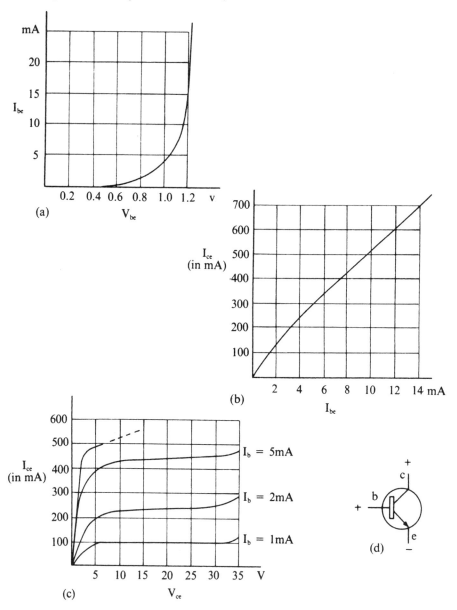

Figure 3.5   NPN silicon transistor characteristics

collector voltage has almost no effect on collector current flow. It is as if the transistor output had a resistance of very high value in series with it.

These graphs show that a transistor connected with its emitter common to both input and output circuits has a low input resistance, fairly large current gain and high output resistance.

Another characteristic which is very useful is the *mutual characteristic*, I_{ce}/V_{be} (or g_m) shown in Figure 3.6 for a typical power transistor. Note the large current

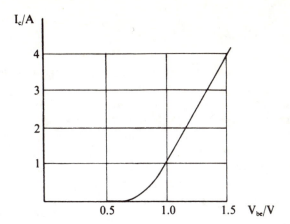

Figure 3.6   The mutual characteristic of a power transistor

values and the nearly straight line of the characteristic.

Although the I_{ce}/V_{be} is often a useful characteristic for an amplifier designer to know, it is not always provided by the transistor manufacturer.

The value, however, does not need to be provided because it depends more on the amount of steady collector current, as set by bias, than anything else. At normal temperatures, around 25°C, the value of g_m for a silicon transistor is 40 × I_c mA/V where I_c is the steady collector current with no signal applied per milliamp of collector current. In other words, for 1mA collector current g_m is 40mA/V, for 2mA collector current, g_m is 80mA/V and so on.

The quantities g_m, h_{fe} and r_{be} are simply related by the formula:

$$r_{be} = \frac{h_{fe}}{g_m}$$

which gives r_{be} in units of kΩ if g_m is in its normal units of mA/V. For example, if a transistor has the h_{fe} value of 100 and is used at a bias current of 1mA, then its g_m value is 40mA/V, and the r_{be} value is $100/40 = 2K5$.

Rules for substitution

When transistors are substituted one for another, the following rules should be obeyed:

1   The substitute transistor must be of the same variety (i.e., silicon, NPN,

switching as opposed to amplifying, etc.)

2 The substitute transistor should have about the same h_{fe} value.

3 The substitute transistor should have the same ratings of maximum voltage and current.

Applications of bipolar transistors

Bipolar transistors are used as current amplifiers, voltage amplifiers, oscillators and switches.

An amplifier, as you know, has two input and two output terminals, but a transistor has only three electrodes. It can therefore only operate as an amplifier if one of its three electrodes is made common to both input and output circuits.

Any one of a transistor's three electrodes can be connected to perform in this common role, so there are three possible configurations: *common emitter*, *common-collector* and *common-base*. The three types of connection are shown in Figure 3.7(a), (b) and (c) respectively.

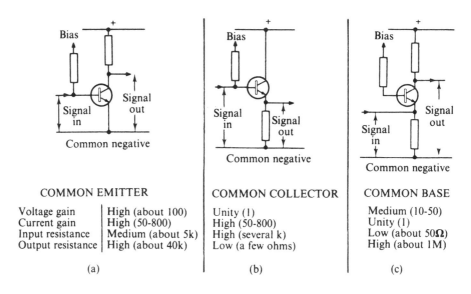

COMMON EMITTER	
Voltage gain	High (about 100)
Current gain	High (50-800)
Input resistance	Medium (about 5k)
Output resistance	High (about 40k)

(a)

COMMON COLLECTOR
Unity (1)
High (50-800)
High (several k)
Low (a few ohms)

(b)

COMMON BASE
Medium (10-50)
Unity (1)
Low (about 50Ω)
High (about 1M)

(c)

Figure 3.7   The three circuit connections of a bipolar transistor

The normal function of a transistor when the base-emitter junction is forward-biased and the base-collector junction reverse-biased is as a current amplifier. Voltage amplification is achieved by connecting a load resistor (or impedance) between the collector lead and the supply voltage (see next chapter). Oscillation is achieved when the transistor is connected as an amplifier with its output fed back, in phase, to its input (Chapter 7). The transistor can also be used as a switch or relay when the base-emitter junction is switched between reverse bias and forward bias.

The three basic bipolar transistor circuit connections are shown in Figure 3.7, with applications and values of typical input and output resistances given below each. Figure 3.7(a) shows the normal amplifying connection used in most transistor circuits. The common-collector connection in Figure 3.7(b), with signal into the base and out from the emitter, is used for matching impedances, since it has a high input impedance and a low output impedance. The common-base connection, with signal into the emitter and out from the collector, shown in Figure 3.7(c) is nowadays used mainly for UHF amplification.

Transistor failure

When transistors fail, the fault is either a short-circuit (s/c) or an open-circuit (o/c) junction; or the failure may possibly be in both junctions at the same time.

An o/c base-emitter junction makes the transistor 'dead', with no current flowing in either the base or the collector circuits. When a base-emitter junction goes o/c, the voltage between the base and emitter may rise higher than the normal 0.6V (silicon) or 0.2V (germanium) – though higher voltage readings are common on fully operational power transistors when large currents are flowing.

An s/c base-emitter junction will allow current to flow easily between these terminals with no voltage drop, but with no current flowing in the collector circuit.

The two above faults are by far the most common, but sometimes a base-collector junction goes s/c, causing current to flow uncontrollably.

All of these faults can be found by voltage readings in a circuit, or by use of the ohmmeter or transistor tester when the transistor is removed from the circuit. (A few types of transistor tester can even be used with the transistor still connected into circuit.)

Summary

Bipolar transistors consist of three regions – emitter, base and collector – with two junctions. Current flows between collector and emitter only when current flows between base and emitter, and when there is current gain (h_{fe}) at least equal to $\frac{I_c}{I_b}$.

The characteristics of a transistor with common-emitter connection show low input resistance, medium output resistance, and values of current gain between 20 and 800 depending on construction.

53

Field-effect transistors

To be strictly accurate, the so-called 'field-effect transistor' is not really a transistor at all. The word 'transistor' is a compression of the term 'transfer resistor', and the FET (as it is commonly abbreviated) does not work like that at all. It is, in fact, more correctly described as a *field-effect device*, because its operation depends on the presence and effects of an electric field. Nevertheless, the term 'field-effect transistor' has become common parlance, so must be used here.

The bipolar transistor relies for its action on making a reverse-biased junction conductive by injecting current carriers (electrons or holes) into it from the other junction. The principles of the field-effect transistor (FET) are entirely different. In any type of FET, a strip of semiconductor material of one type is made either more or less conductive because of the presence of an electric field pushing carriers into the semiconductor or pulling them away.

There are two types of field-effect transistor – the *junction FET* and the *metal-oxide silicon FET*, or *MOSFET*. Both work by controlling the flow of current carriers in a narrow channel of silicon. The main difference between them lies in the method by which the flow is controlled.

Look first at the structure of the junction FET (Figure 3.8). A tiny bar of silicon

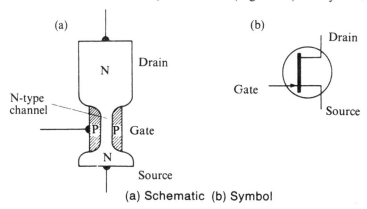

(a) Schematic (b) Symbol

Figure 3.8   The junction FET

of either type (the N-type is actually illustrated) has a junction formed near one end. Connections are made to each end of the bar, and also to the P-type material at the junction. The P-type connection is called the *gate*, the end of the bar nearest the gate the *source*, and the other end of the bar the *drain*.

A junction FET of the type illustrated is normally used with the junction reverse-biased, so that few moving carriers are present in the neighbourhood of the junction.

The junction, however, forms part of the silicon bar, so that if there are few carriers present around the junction, the bar itself will be a poor conductor. With less reverse bias on the junction, a few more carriers will enter the junction and the silicon bar will conduct better; and so on as the amount of reverse bias on the junction decreases.

When the voltage is connected between the source and the drain, therefore, the amount of current flowing between them depends on the amount of reverse bias on the gate; and the ratio: $\dfrac{\text{Source-drain current}}{\text{Gate voltage}}$ is called the *mutual conductance*, whose symbol is g_m. This quantity, g_m, is a measure of the effectiveness of the FET as an amplifier of current flow.

For most FETs, g_m values are very low, only about 1.2 to 3 mA/V, as compared with corresponding values for a bipolar transistor of from 40mA/V (at 1mA current) to several ampere/volts at high rates of current flow. Because the gate is reverse-biased, however, practically no gate current flows, so that the resistance between gate and source is high – very much higher than the resistance between base and emitter of a working bipolar transistor.

Figure 3.9 shows the basic construction of the metal-oxide-silicon FET. A

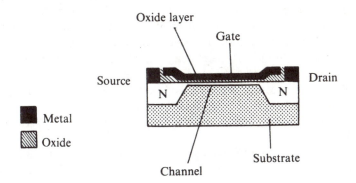

Figure 3.9   Construction of MOSFET

silicon layer, called the *substrate*, is used as a foundation on which the FET is constructed. The substrate may have a separate electrical connection, but it takes no part in the FET action and is usually connected either to the source or the drain. Two regions which are both doped in the opposite polarity to the substrate are then laid on the substrate and joined by a thin channel. In the illustration, Figure 3.9, the substrate is of P-type silicon and the source, drain and channel are of N-type so that there is a conducting path between the N-type source and drain regions.

The gate is insulated from the channel by a thin film of silicon oxide, obtained by oxidizing some of the silicon of the channel, and a metal film is deposited over this insulating layer to form the gate itself.

A positive voltage applied to the gate has the effect of attracting more electrons into the channel, and so increasing its conductivity. A negative potential so applied would repel electrons from the channel and so *reduce* its conductivity.

Both N-channel and P-channel devices can be made. In addition, the channel can be either doped or undoped (or very lightly doped). If the channel is strongly doped there will be a conducting path of fairly low resistance between the source and the drain when no bias is applied to the gate. Such a device is usually operated with a bias on the gate that will reduce the source-drain current, and is said to be used in *depletion mode*. When the channel is formed from lightly-doped or undoped material it is normally non-conducting, and its conductivity is increased by applying bias to the gate in the correct polarity, using the FET in *enhancement mode*. Enhancement mode is more common.

With the gate-to-source voltage equal to zero, the device is cut off. When a gate voltage that is positive with respect to the channel is applied, an electric field is set up that attracts electrons towards the oxide layer. These now form an induced channel to support a current flow. An increase in this positive gate voltage will cause the drain-to-source current flow to rise.

Junction FETs cause few handling problems provided that the maximum rated voltages and currents are not exceeded. MOSFETs, on the other hand, need to be handled with great care because the gate must be completely insulated from the other two electrodes by the thin film of silicon oxide. This insulation will break down at a voltage of 20 to 100V, depending on the thickness of the oxide film. When it does break down, the transistor is destroyed.

Any insulating material which has rubbed against another material can carry voltages of many thousands of volts; and lesser electrostatic voltages are often present on human fingers. There is also the danger of induced voltages from the a.c. mains supply.

Voltages of this type cause no damage to bipolar transistors or junction FETs because these devices have enough leakage resistance to discharge the voltage harmlessly. The high resistance of the MOSFET gate, however, ensures that electrostatic voltages cannot be discharged in this way, so that damage to the gate of a MOSFET is always possible.

To avoid such damage, all MOS gates that are connected to external pins are protected by diodes which are created as part of the FET during manufacture and which have a relatively low reverse breakdown voltage. These protecting diodes will conduct if a voltage at a gate terminal becomes too high or too low compared to the source or drain voltage level, so avoiding breakdown of the insulation of the gate.

The use of protective diodes makes the risk of electrostatic damage very slight for modern MOS devices, and there is never any risk of damage to a gate that is connected through a resistor to a source or drain unless excessive d.c. or signal voltages are applied. Nevertheless, it is advisable to take precautions against electrostatic damage, particularly in dry conditions and in places where artificial fibres and plastics are used extensively. These precautions are:

1   Always keep new MOSFETs with conductive plastic foam wrapped round
    their leads until after they have been soldered in place.
2   Always short the leads of a MOSFET together before unsoldering it.
3   Never touch MOSFET leads with your fingers.
4   Never plug a MOSFET into a holder when the circuit is switched on.

Exercise 3.3

Using a 2N3819 junction FET, connect the circuit shown in Figure 3.10. The

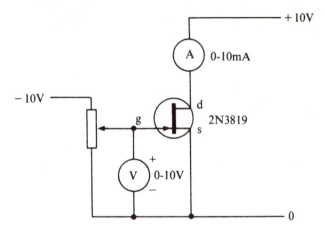

Figure 3.10   Circuit for Exercise 3.3

milliammeter measures the current through the channel, and the voltmeter measures the (negative) bias on the gate.

   For the bias voltages listed, fill in the readings of current. Find also the *cut-off
voltage*, which is the gate voltage at which channel current flow just reaches zero.

   FET's can be used in circuits similar to those in which bipolar transistors are
used, but they give low voltage gain and are only used when their peculiar
advantages are required. These are:

1   FET's have a very high input resistance at the gate – a useful feature in
    voltmeter amplifiers.
2   FET's perform very well as switches, with channel resistance switching
    between a few hundred ohms and several megohms as gate voltage is varied.
3   The graph of channel current I_{ds} plotted against V_{gs} – the voltage between the
    gate and the source – is noticeably curved in a shape called *a square law* (see
    Figure 3.11). This type of characteristic is particularly useful for signal mixers
    in superheterodyne receivers.

   Double-gate MOSFET's are used as mixers and as RF amplifiers in FM

57

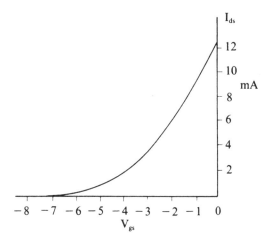

Figure 3.11   A typical FET characteristic

receivers. The shape of their characteristic also gives less distortion in power amplifiers, and high-power FET's are now available for use in high-quality audio equipment.

FET and MOSFET failure

Failure of a junction FET can be caused by either an open-circuit or a short-circuit junction. MOSFET failure is almost always caused by breakdown of the insulating silicon oxide layer. In either case, gate voltage can no longer control current flow in the channel between source and drain, and pinch-off becomes impossible.

If very large currents have been allowed to flow between source and drain, the channel may burn out to an open circuit.

Summary

FETs are devices that depend on a junction action different to that of bipolar transistors. Junction FETs are usually operated with their single junction reverse-biased (i.e., in the depletion mode).

MOSFETs have almost infinite gate resistance, and the leads must not be touched unless the gate is first shorted to the other two electrodes.

FETs are used in applications where their high input resistance, good switching characteristics and low-noise factor outweigh their poor voltage gain.

The unijunction

The unijunction has a construction rather similar to that of a junction FET, but with the junction formed about midway along the silicon bar. When the bar is connected across a supply voltage (with the voltage at Base 1 always lower than that at Base 2), the resistance of the bar will cause potential divider action, so that the voltage of the N-part of the junction (assuming that an N-channel is used) will be somewhere between zero volts and supply voltage, according to the position of the junction.

If the channel is only lightly doped, the resistance will be high, and very little current will flow when a voltage is placed across the channel.

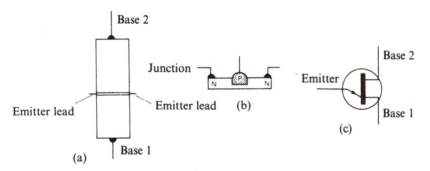

Figure 3.12 **The unijunction**
(a) Schematic (b) In profile (c) Symbol

When the emitter terminal is at the same voltage as the Base 1 terminal (which is always, it will be remembered, lower than B2), the junction is reverse-biased because the part of the junction lying in the channel is at a higher voltage than the emitter.

If emitter voltage be now raised slowly, current will start to flow across the junction when emitter voltage is about 0.5V above the voltage at the N-part of the junction (again assuming an N-channel unijunction).

When this happens, carriers (electrons in this case) enter the channel, making it a good conductor. Current can now pass freely between emitter and Base 1, and also between B2 and B1. This conductivity continues until emitter voltage is again reduced to almost zero voltage.

The unijunction is used to generate trigger pulses from slowly-changing waveforms, or as an oscillator generating sawtooth waveforms or pulses. It is mainly used in the circuits used to trigger thyristors (see below).

The voltage needed at the emitter for the unijunction to start conducting is called the *striking voltage*. The ratio of the values: $\dfrac{\text{Striking voltage}}{\text{Supply voltage}}$ is called the *intrinsic stand-off ratio*, and is a constant in a given unijunction because it depends only on the position of the junction along the bar.

For example, if a unijunction supplied with 6V between B1 and B2 triggers on 4V, then its stand-off ratio is 4/6, equal to 2/3 or 0.67. When 12V is connected between B1 and B2, the trigger voltage will be $2/3 \times 12 = 8V$.

When a unijunction has to be replaced, the replacement must possess the same values of maximum voltage and current, and the same intrinsic stand-off ratio.

Exercise 3.4

Connect the circuit of Figure 3.13, and measure the supply voltage E.

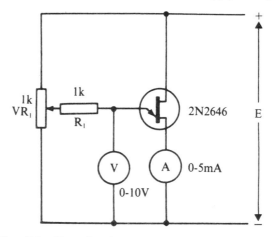

Figure 3.13   Circuit for Exercise 3.4

Starting at zero volts, measure the emitter voltage and raise its value slowly until the unijunction triggers, causing the milliammeter suddenly to give a reading. Note the reading of the voltmeter just before the triggering, and calculate the value of the intrinsic stand-off ratio.

Now change the value of the supply voltage and repeat your readings. Is the same value of intrinsic stand-off ratio obtained with the new supply voltage?

The programmable unijunction transistor (PUT) is a form of unijunction which allows the firing voltage (and the intrinsic stand-off ratio) to be controlled by a potential divider. The 2N6027 is a typical example. Another device, the silicon-controlled switch (SCS) is a four-layer semiconductor which can be connected either as a PUT or as a thyristor (see later).

The thyristor

The thyristor is a switching device utilizing four doped layers whose symbol and schematic construction are shown in Figure 3.14.

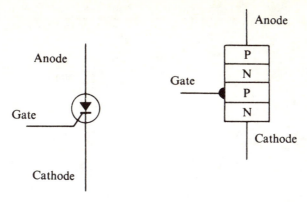

Figure 3.14   The thyristor, symbol and schematic construction

When gate voltage is zero, the device offers high resistance irrespective of the polarity of the anode-to-cathode voltage, because there is always one PN junction with reverse bias. Even when a thyristor is forward-biased, with the anode positive and the cathode negative, it will only be driven into conduction when a short positive pulse is applied to its gate.

Once it has been switched on in this way, the thyristor will remain conducting until either:

(a) the voltage between anode and cathode falls to a small fraction of a volt; or
(b) current flow between anode and cathode falls to a very low value.

The important ratings for a thyristor are its maximum average current flow, its peak inverse voltage, and its values of gate-firing voltage and current. A small thyristor will fire at a very small value of gate current, but a large one calls for considerably greater firing current ($100\mu A$ or more).

Exercise 3.5

Connect the circuit of Figure 3.15. Any small 1-ampere thyristor such as BTX18-400, TICP106D will do for the job. The purpose of the lamp bulb is to indicate when the thyristor has switched on.

Use a d.c. supply as shown, and measure voltage and current at the gate just as the thyristor fires – noting also that both values change after the thyristor has fired.

After the lamp lights, disconnect the gate circuit, and note that this has no effect.

Switch off the power supply, and then switch on again. The lamp will not now light because the thyristor has switched off.

Repeat the operation, this time using an unsmoothed supply. The thyristor

61

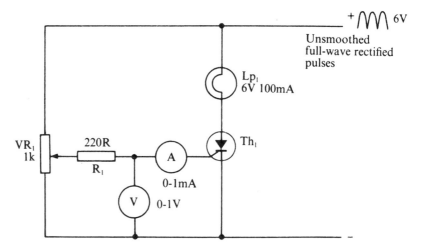

Figure 3.15   Circuit for Exercise 3.5

now switches off when the gate voltage is switched off because the unsmoothed supply reaches zero voltage 100 times per second (assuming a full-wave rectified 50Hz supply).

Thyristors are used for power control, using either a.c. or an unsmoothed rectified supply. By being made to fire at different parts of the cycle, the thyristor can be made to conduct for different percentages of the cycle, thus controlling average current flow through the load.

An alternative method used in furnace control allows the thyristor to conduct for several cycles of a.c., and then to cut off the current flow for several more cycles. By altering the ratio: $\dfrac{\text{Cycles conducting}}{\text{Cycles non-conducting}}$, average power can be controlled.

There is one form of thyristor, the gate-turn-off thyristor (GTO thyristor) which is triggered on by a positive pulse of fairly high current (typically 100mA) and triggered off by a negative pulse of lower current; both pulses applied to the gate.

Exercise 3.6

Connect the thyristor circuit of Figure 3.16 and switch on. Use the potentiometer to control lamp brightness by altering the time in the cycle at which the thyristor fires.

Note that an unsmoothed full-wave rectified supply is essential.

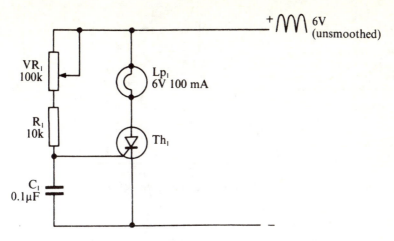

Figure 3.16   Circuit for Exercise 3.6

Thyristor failure

Failure of a thyristor can be caused by an open-circuit gate, or by internal short-circuits.

When the gate is open-circuit, the thyristor will fail to conduct at any gate voltage. When an internal short circuit is present, the thyristor acts like a diode, conducting whenever the anode is more than 0.6V positive to the cathode, so making control impossible.

A completely short-circuit thyristor is able to conduct in either direction, with similar total loss of control.

Most thyristor circuits include inductors in series with the anode, in order to suppress the radio-frequency interference which is caused by the sudden switch-on of the thyristor.

The diac

A diac is a triggering diode which is often used in the gate circuit of a thyristor or triac (see below).

At low voltages, a diac will not conduct in either direction, but at a triggering voltage of either polarity it becomes a good conductor in both directions.

When a diac is used, the gate of the thyristor can be connected to the cathode by a low-value resistor to avoid accidental triggering. In addition, if the triggering waveform rises slowly, the diac will ensure that the thyristor is switched on by a sharp pulse of current, so avoiding uncertain triggering times.

The symbol for, and a typical use of, the diac are shown in Figure 3.17.

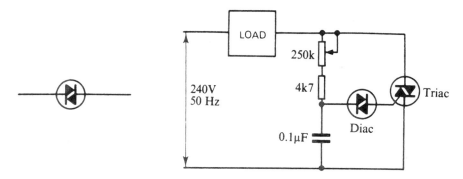

Figure 3.17   The diac, symbol and typical connection

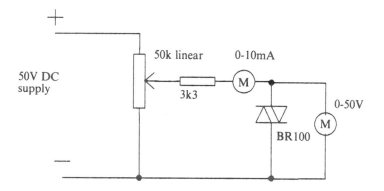

Figure 3.18   Circuit for Exercise 3.7

Exercise 3.7

Connect a diac, such as the BR100, into the circuit of Figure 3.18. Ensure that the potentiometer is set so as to provide zero voltage at the output when the power is applied, and then switch on. Read values of V and I for each 5V step of voltage until current starts to flow, and then on each 2mA change of current up to 10mA. Now return the potentiometer setting to the zero-output position and reverse the polarity of the power supply. Repeat the readings. Plot a graph of V against I, allowing for both positive and negative values.

The triac

A triac is a two-way thyristor which, when triggered, will conduct in either direction. The terminals are B1, B2 and gate (the terms 'anode' and 'cathode' cannot be used because the current can flow in either direction). The gate can be triggered by a pulse of either polarity, but the most reliable triggering is achieved when the gate is pulsed positive with respect to B1.

In most triac circuits, B1 and B2 have alternating voltages applied to them, so that the gate must receive the same waveform as B1 when it is not being triggered.

To trigger the gate, a pulse must be added to the waveform already present, and this is most easily done by using a pulse transformer driven by a trigger circuit.

Figure 3.19 shows the symbol for a triac, and part of a typical triggering circuit.

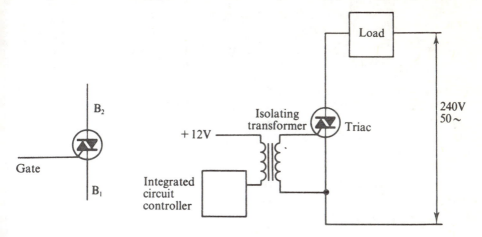

Figure 3.19   The triac, symbol and typical connection

A transformer is used to isolate the triac, which works at line voltage, from the low-voltage control circuit.

The causes of failure in triacs are the same as they are for thyristors.

The advantages of using thyristors and triacs for power control are:

1   Both thyristor and triac are either completely off, with no current flowing, or fully on, with only a small voltage between the terminals. Either way, very little power is dissipated in the semiconductor.
2   Operation can be either at line or at higher frequencies – unlike relays or similar electromechanical switches.
3   Thyristors and triacs are *self-latching* – which means that they stay conducting once they have been triggered. A relay, by contrast, needs a current passed continuously through its coil to keep it switched over.

Exercise 3.8

HAZARD: High voltages. This circuit *must* be supplied from an isolating transformer, and all live parts of the circuit must be covered to protect against accidental contact.

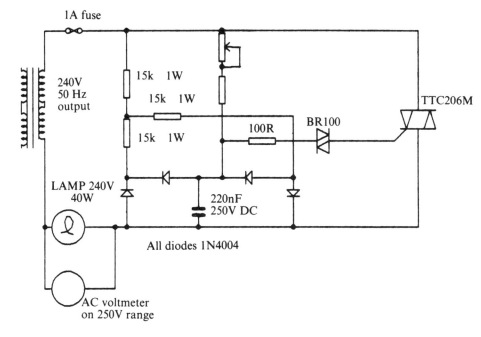

1A fuse

240V
50 Hz
output

15k 1W

15k 1W

15k 1W

100R

BR100

TTC206M

LAMP 240V
40W

220nF
250V DC

All diodes 1N4004

AC voltmeter
on 250V range

Figure 3.20   Circuit for Exercise 3.8

Figure 3.20 shows a circuit using a triac to control the power to a 40W lamp. The potentiometer adjusts the charging current to the capacitor, so determining at what part in each half-wave the voltage across the diac will be enough to cause conduction and so fire the triac. Note how the voltage registered on the a.c. voltmeter varies as the potentiometer is adjusted.

For further work, inspect the waveforms in the circuit across the capacitor and on each side of the diac. An oscilloscope can be used since the circuit is fed from an isolating transformer, but remember that an oscilloscope *must never be used to check waveforms in a triac or thyristor circuit which is directly connected to the mains.*

Magnetic-dependent semiconductors

When a current-carrying conductor is exposed to a magnetic field, a mechanical reaction occurs which tends to cause the conductor to move. This is the principle of the electric motor which was described in Volume 1. The direction of the motion can be predicted from Fleming's Left-Hand Rule for motors.

If the conductor is *prevented* from moving, however, the current carriers flowing within the conductor are deflected to one side. The result is that a voltage is developed across the width of the conductor itself.

In normal conductors this effect is of no practical importance; but when *a magnetic field* is applied to semiconductor materials, the voltage so generated becomes significant. It is put to practical use in a group of semiconductors called *Hall-effect devices*, illustrated in schematic form in Figure 3.21 below.

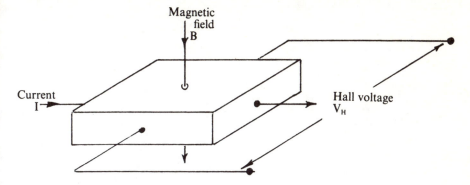

Figure 3.21   Schematic diagram of a Hall-effect device

A Hall-effect device consists of a very small (about 2mm × 2mm × 0.5mm) slab of gallium arsenide, indium antimonide or silicon on to which two pairs of electrodes are evaporated, as shown in Figure 3.21. A control current I and a magnetic field B are applied as shown, and the Hall voltage V_H is developed at right angles to both of them. The value of V_H is proportional to both the current I and the strength of the magnetic field B, the constant of proportionality depending on the dimensions and nature of the semiconductor material.

Given a constant current, the Hall voltage is directly proportional to the strength of the magnetic field. The first and most obvious application for the device is therefore the measurement of the field strengths of both permanent magnets and electro-magnets. But it is also used in conjunction with switching transistors in d.c. electric motors, where it can replace the mechanical brush gear which is so prone to wear and to the generation of radio interference by arcing. The device can also be used in contact-less switches, where the movement of a permanent magnet sets up the Hall voltage, which in turn triggers a semiconductor switch.

Magnetic-dependent resistors

When the application of the magnetic field in the Hall-effect device deflects the current carriers to one side of the conductor, they are caused to flow through a smaller cross-sectional area. This increases the effective resistance of the device, and so reduces current flow. The characteristic of a typical magnetic-dependent resistor is shown in Figure 3.22.

Over a normal range of magnetic fields strengths, device resistance can in this way be made to vary by a factor of about 5.

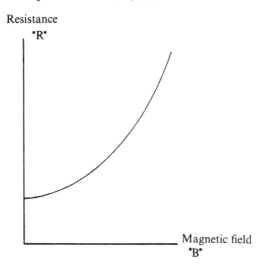

Resistance
'R'

Magnetic field
'B'

Figure 3.22   Characteristic of a typical magnetic-dependent resistor

One particularly useful application of a magnetic-dependent resistor is in the 'clamp-on' type of current meter used to measure current flow in power supplies drawing power directly from the electrical mains.

Summary

Unijunctions, thyristors, diacs and triacs are all switching devices which switch suddenly from a non-conducting state to a conducting state.

Unijunctions and diacs are used to provide trigger pulses for thyristors and triacs, which in turn are used to control power in a load.

The supply for thyristor circuits is usually unsmoothed rectified current. Triacs operate from an a.c. supply.

Hall-effect devices and magnetic-dependent resistors are used for specialized purposes such as sensing magnetic fields or in contactless switching.

Introducing integrated circuits

The actual size of the silicon chip in a transistor is very small compared to the size of the transistor case; and the size of the active part of the chip is smaller still.

All silicon transistors are manufactured by processes which closely resemble the printing and etching techniques that are used for the construction of PCBs. A pattern of contacts and connections can be made in the form of a metal mask which can be placed over a silicon chip, so that operations such as oxidation (creating an insulating surface), doping, or metallizing (to make external contacts) can be carried out in selected areas.

If a suitable set of masks can be made, forming 100 or more transistors on to a chip is as easy as forming one, and no additional manufacturing steps are needed. Since the amount of doping of silicon controls its resistivity, it is possible to use another mask to form resistive channels onto the surface of a chip which will connect between transistors that have already been formed. With further masking operations, small-value capacitors can be formed, using oxidation to make the insulation and metal layers to form one or more plates. Oxidized layers can also be used to insulate channels which cross over each other and to allow one layer to be built on top of another.

Using these and later techniques, complete circuits can be built up on the tiny space of a silicon chip, and when larger chip surfaces are used the number of components can be very large indeed. Because transistors, particularly MOS transistors, are the easiest components to fabricate and connect in this way, much ingenuity has gone into designing circuits which make use of transistors other than capacitors or resistors. Where the use of capacitors, resistors or inductors is essential, these components can be connected externally.

ICs are classed by 'scale of integration', meaning the number of transistors (strictly speaking, the number of gate circuits) on each chip. Early ICs were all of small-scale integration (SSI) with 10–20 transistors per chip, but the technology advanced very rapidly to medium-scale (MSI) with about 100–200 transistors per chip and then to large scale (LSI) with 1000 or more transistors per chip. Many of the chips that are now used in computers are VLSI, meaning 10,000 or more transistors per chip, and chips containing 1,000,000 or more transistors are being manufactured. The LSI and VLSI chips use MOS techniques exclusively, and are manufactured by computer-controlled methods.

The advance of ICs has made the use of individual transistors (discrete circuitry) almost unknown in digital circuits and rare in analogue circuits. It is still important to know the fundamentals of transistor action, partly because transistors will still be found in equipment that will need servicing and partly because the transistor is the building-block of the IC. This book will not neglect transistor circuits, but will necessarily concentrate on the use of ICs and how faults can be diagnosed and repaired in circuits which use ICs. In many modern circuits, the only discrete devices found are bridge rectifiers, thyristors and other power-handling semiconductors, and high-power transistors.

The advantages of using ICs are, in decreasing order of importance:

1   Greatly increased reliability. A circuit with several thousand components suffers from reliability problems because the chances of one component failing become greater as the number of components and their interconnections is increased. The IC, by contrast, is a single component which replaces all these discrete components.
2   More predictable performance because all ICs use more elaborate circuits.
3   Greatly reduced cost because a complete circuit can be made, once design

costs are paid for, for much the same price as a single semiconductor.

4   Greatly reduced size, allowing a huge amount of circuitry to be put into a small space.

ICs can be soldered into normal printed circuit boards; but for some military and industrial uses, thick-film circuits are used. These are like miniature printed-circuit boards, only with printed silicon chips and other miniature components soldered or welded in place, and resistors or capacitors formed by the metallic films on the boards.

ICs can be obtained in packages which can be surface-mounted, reducing both stray capacitance and stray inductance of connections. The differences between thick-film and thin-film (IC) circuits are:

1   Thin-film ICs are formed on silicon and contain transistors, resistors and capacitors formed in silicon.

2   Thick films are made of a plate of conductive metal such as nickel, stuck or welded to a material such as glass or sapphire, with transistors or IC chips attached.

3   Thin-film ICs are usually made so that they can be used in a great variety of applications, while thick-film circuits are more likely to be 'dedicated' – or made for one purpose only.

Handling ICs

ICs may be either bipolar or metal-oxide-silicon; and the handling precautions which must be taken with MOS transistors apply equally to MOS ICs.

The maximum working voltages for ICs are generally lower than they are for bipolar transistors: and maximum power dissipation is usually low except in the case of specially-designed power output ICs. In some cases, supply voltages must be carefully regulated. For example, ICs of the logic TTL family need a 5V supply which must never fall below 4.8V nor rise above 5.4V.

Other types need balanced power supplies, such as ± 9V, which have to be provided from suitable rectifier circuits.

The most common form of IC package is the *DIL*, a rectangular slab with two lines of pins sticking out on either side of it. The name is composed of the initials of the words *dual-in-line*, referring to the two rows of pins. Pin numbers in common use are 8, 14, 16 and 24 – with a few large ICs (mainly designed for computer use) having greater numbers of pins.

Pin spacings are always multiples of 2.54mm (0.1″) apart (this is referred to as a *2.5mm matrix*), so that they fit exactly into the standard spacing of the holes in printed circuit boards. Great care must be taken when inserting pins to avoid bending them in any direction.

The pin-numbering system shown in Figure 3.23 always starts from the *identi-*

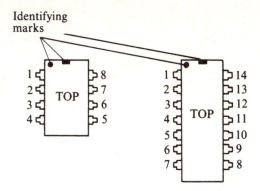

Figure 3.23   Pin-numbering systems for DIL integrated circuits

fying mark and runs in the way illustrated, even for DIL ICs containing many more pins than the 8-pin and the 14-pin ICs depicted.

Some modern ICs, notably microprocessors, require more than 40 pins, and the DIL type of construction becomes unsuitable, with quite long internal leads from the chip itself to the pins near the corners. Several of these larger chips make use of a pin-grid casing, such as the type illustrated in Figure 3.24 with 68 pins arranged in a double decker format. The no. 1 pin is marked by a larger wedge taken from its corner.

Most ICs cannot be easily tested before being soldered into a circuit, so it is important that they be bought from reputable sources. The circuit into which the IC is to be soldered should be checked carefully for open circuits or accidental short circuits. Note that, with printed circuit tracks only 2.5 mm apart, small splashes of solder can easily cause a short between tracks.

When an IC is soldered into place, its pin no. 1 should be soldered first, then the circuit connections checked. Its opposite pin is then soldered in, and the circuit checked again to make quite sure that the IC has been put in the right way round. If the connections are incorrect, it is possible to remove the IC quite easily at this stage.

If the circuit seems satisfactory, the remaining pins are then soldered in.

If a faulty IC has to be removed from a circuit – and the job should not be attempted until it is quite certain that the IC really *is* faulty – the easiest method is to cut off all the pins, remove the body of the IC, then unsolder each pin, pulling the pin out with tweezers while the soldering iron is used to melt the solder. When all this has been done, there is at least no chance that the faulty IC will ever again be mistaken for a good one!

Alternative methods which preserve the pins for future use employ solder-removing tools such as solderwick or solder pumps. By these methods the solder is removed from the pins so that the IC can be lifted out intact. This method should be adopted if there is any doubt whether it is the IC or the circuit itself

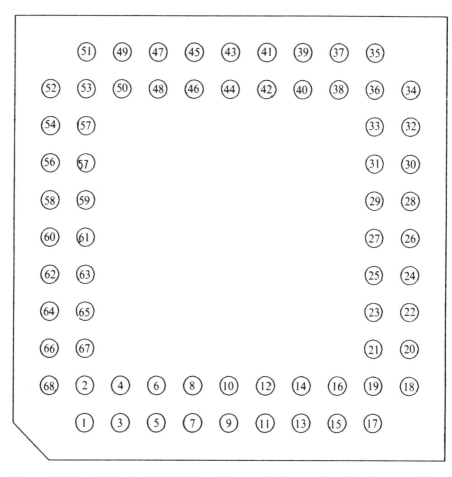

Figure 3.24 A 68-pin PGA (Pin-grid array) form of packaging used for microprocessor and other VLSI ICs

which is at fault.

If the IC is of the MOS type, soft copper wire should always be wrapped around the pins before de-soldering begins.

When there is an IC failure, it is usually fairly simple to find out which IC has failed, but less easy to find out why. Every IC will have some d.c. bias voltage which can be measured. If bias voltages are correct, and if the correct input signals are being applied, then the correct output signals must appear if the IC is operating correctly.

The exceptions to this rule are some power-output or stabilizer ICs in which the temperature becomes too high. If a fault is suspected in an IC of this type, testing should be deferred until the IC has cooled.

In many respects, the servicing of ICs is rather easier than is the servicing of discrete transistor equipment.

Summary

ICs are etched on to thin silicon plates in the same way as are transistors, but in such a way that a complete circuit is made in one manufacturing process.

Thick film circuits are manufactured from metal film deposited on insulators, with resistors and capacitors formed in place. The active components (transistors, ICs, etc.) have to be added later.

Multiple-choice test questions

1 Which transistor in Figure 3.25 is most likely to be conducting?

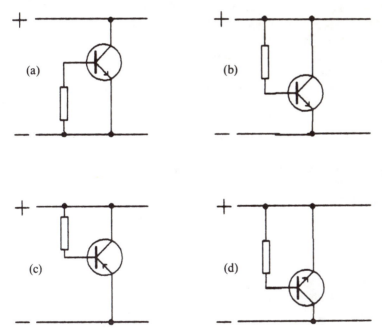

Figure 3.25

2 The table shows the results of resistance readings, shown as H for high or L for low, made on three pins X, Y and Z of an unknown NPN transistor.

X+ to Y−	X− to Y+	Y+ to Z−	Y− to Z+	X+ to Z−	X− to Z+
H	L	L	H	H	H

Each column shows the terminals that are tested and the true polarity of the connections. The transistor connections are:
(a) X=e Y=b Z=c
(b) X=b Y=c Z=e
(c) X=c Y=b Z=e
(d) X=e Y=c Z=b.

3   A transistor has g_m = 50 and h_{fe} = 500. Its base-emitter input resistance will be:
(a) 1k
(b) 10k
(c) 1M
(d) 10M.

4   An N-channel MOSFET is to be used in a circuit in enhancement mode. In order to ensure that current flows in the channel, the FET
(a) must have its gate connected to its source
(b) must have its gate open-circuit
(c) must have the gate at a negative bias relative to the source
(d) must have the gate at a positive bias relative to the source.

5   A thyristor is used to switch a smoothed d.c. supply. Once the thyristor has been switched on, it can be turned off by:
(a) pulsing the gate negative
(b) momentarily short-circuiting the supply
(c) pulsing the gate positive
(d) applying AC to the gate.

6   The internal circuitry of ICs is designed so as to
(a) make maximum use of resistors
(b) make maximum use of capacitors
(c) make maximum use of transistors
(d) make maximum use of inductors.

4   Voltage amplifiers

Summary

Amplifying characteristics, linear amplification, transfer characteristic. Mutual characteristics, output characteristics. Bias. Gain and Bandwidth. Coupling methods. Negative feedback. Differential amplifiers. Operational amplifiers.

Basic parameters

The bipolar transistor and the FET both control the current flow at their output terminals (collector and drain respectively); and in both cases this current at the output can be controlled by the voltage at the input.

The ratio:

$$\frac{\text{Change of current at output}}{\text{Change of voltage at input}} \text{ (given a constant supply voltage)}$$

is called the *mutual conductance* of the particular bipolar transistor or FET to which it applies. Its symbol is g_m. The values of mutual conductance obtainable from bipolar transistors are much greater than are those from FETs. In the mutual conductance graph shown in Figure 4.1, for instance, a 30mV input wave gives a 1mA current flow at the output. The mutual conductance g_m is therefore $\frac{1}{0.03} = 33.3\text{mA/V}$.

As was noted in Chapter 3, the value of g_m depends mainly on the steady

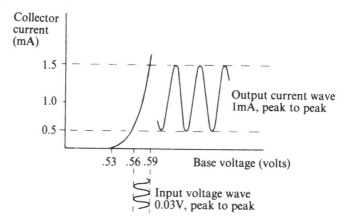

Figure 4.1   Mutual conductance graph

value of collector current, and is approximately $40 \times I_c$ mA/V for a silicon transistor.

The method of operation is as follows. A signal voltage, alternating from one peak of voltage to the other, at the input produces a signal current, alternating from one peak of current to the other, at the output. To convert this signal current into a signal voltage again, a load is connected between the output terminal and the supply voltage. In a d.c. or an audio amplifier, a resistor can be used as the load; but for i.f. or r.f. amplifier circuits a tuned circuit (which behaves like a resistor at its tuned frequency) is used instead.

When the device in question is connected in the common-emitter mode, the use of a load resistor causes the output voltage to be inverted compared to the input voltage. A higher steady voltage at the input causes a greater current flow at the output. A larger voltage is therefore dropped across the load resistor, which causes the output voltage to be lower.

Any amplifying stage can give current gain, voltage gain or power gain – the amounts of gain which can be obtained depending on the way the circuits are connected. Figure 4.2 shows the three possible amplifying connections of a single transistor, with their relative gain values (bias components not shown).

Note that circuits using bipolar or field effect transistors give power gain – the amount of this gain is calculated by multiplying voltage gain by current gain.

The transfer characteristic

The behaviour of an amplifier can be clearly read from a graph of its output voltage or current plotted against its input voltage or current (for given values of load resistance and supply voltage). For example, the transfer characteristic of a small bipolar transistor is shown in Figure 4.3. An input current wave of

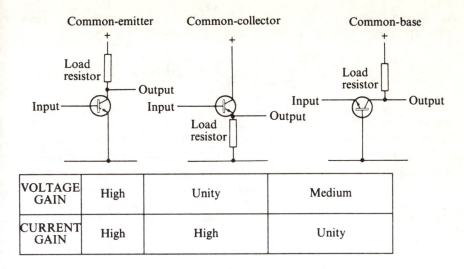

	Common-emitter	Common-collector	Common-base
VOLTAGE GAIN	High	Unity	Medium
CURRENT GAIN	High	High	Unity

Figure 4.2   Amplifying connections of single transistor

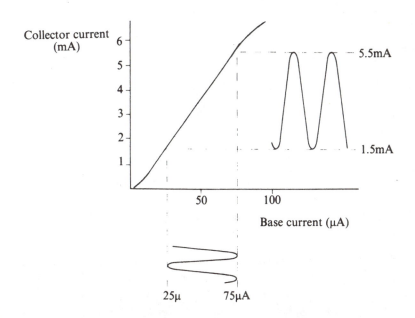

Figure 4.3   Transfer characteristic of small bipolar transistor

50μA peak-to-peak produces an output current wave of 4mA peak-to-peak. The current gain h_{fe} of the transistor under these conditions is thus

$$\frac{\text{Collector current swing}}{\text{Base current swing}} = \frac{4\text{mA}}{50\mu\text{A}} = 80.$$

The voltage wave produced is in this case an exact copy of the input wave, because that part of the graph which represents the values of voltage and current is a straight line. Because a straight-line characteristic produces a perfect copy, such an amplifier is called a *linear amplifier*. If the input signal had been greater, with peak values 0 and 30μA, the output signal would not have been a perfect copy of the input wave because the output voltage cannot exceed 6V even when the input current changes from 17μA to 30μA. The output signal would therefore be distorted because the part of the characteristic now being used would not be a section of a straight line.

Figure 4.4 shows the plots of the output current/input voltage characteristics

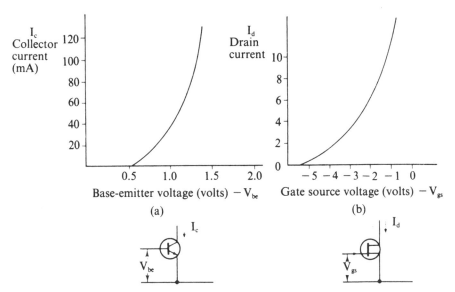

Figure 4.4 The mutual characteristics of (a) a typical bipolar transistor, and (b) a typical junction FET

of (a) a bipolar transistor and (b) an FET. The shape of these characteristics shows that reasonably good linear amplification is possible only if a small part of the characteristic is selected for use. It is not, for example, possible to use an input voltage of less than 0.6V for the bipolar transistor.

In practice, however, these two latter transfer characteristics are seldom drawn – the more usual type of information given being the *output characteris-*

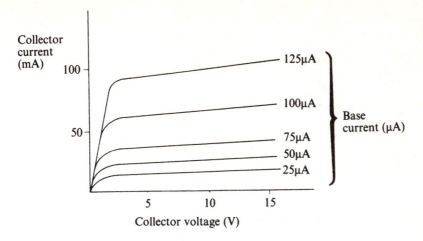

Figure 4.5   Output characteristic for a typical bipolar transistor

tic. In Figure 4.5 every line is the graph of I_c/V_c for a given base current I_b. The difference in spacing shows that amplification will be non-linear. In other words, if the output characteristic lines are drawn for equal changes of input voltage or current, the unequal spacing of the lines indicates that the transfer characteristic must be curved, producing the effects already noted.

Bias

The mutual characteristics of the bipolar transistor shown in Figure 4.4(a) make it clear that if such a transistor is to be used as a linear amplifier, the output current must never cut off nor must the output voltage be allowed to reach zero (its so-called *bottomed condition*).

Ideally, when no signal input is applied, output current should be exactly half-way between these two conditions. This can be achieved by *biasing* – i.e., by supplying a steady d.c. input which ensures a correct level of current flow at the output. A correctly-biased amplifier will always deliver a larger undistorted signal than will an incorrectly-biased one.

A correctly-biased amplifier operating in such conditions, with the output current flowing for the whole of the input cycle, is said to be operating under *Class A conditions*.

Exercise 4.1

Connect the circuit shown in Figure 4.6, in which Tr1 can be any medium-current silicon NPN transistor such as 2N3053, 2N1711, 2N2219 or BFY60. Set

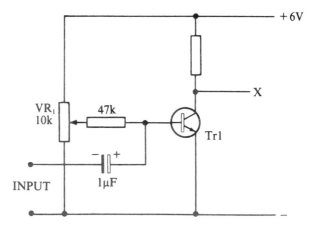

Figure 4.6   Circuit for Exercise 4.1

the signal generator to deliver a signal of 50mV peak-to-peak at 1kHz when connected to the amplifier input. Connect the oscilloscope to the output terminals, with the Y-input of the oscilloscope to Point X. Set the oscilloscope Y-input to 1V/cm and the timebase control to 1 ms/cm, and switch on the oscilloscope. When the trace is visible, adjust the potentiometer VR_1 to its minimum voltage position, and switch on the amplifier circuit.

Note that there is no output from the amplifier because it is incorrectly biased. Gradually increase the bias voltage by adjusting VR_1 until a waveform trace appears. Draw the wave-shape.

Continue to adjust the bias voltage until the waveform seen on the oscilloscope screen is a pure sine-wave – it may be necessary to adjust the amplitude of the input wave to achieve this.

Then increase the bias still further until distortion becomes noticeable again, and sketch this waveform also. You will see that with too little bias the amplifier cuts off, causing the top of the waveform to flatten. With too much bias, the amplifier bottoms, causing the bottom of the waveform to flatten.

Summary

All transistor-type amplifier circuits produce a gain in power, and both voltage gain and current gain can be obtained.

Gain is obtained by generating an output waveform under the control of the input waveform, and the transfer characteristic is the graph of output plotted against input.

If the graph of the transfer characteristic is a straight line for the quantities being plotted (power, voltage or current), the amplifier is *a linear amplifier of that quantity*, and the output wave will be a perfect copy of the input wave, with no distortion.

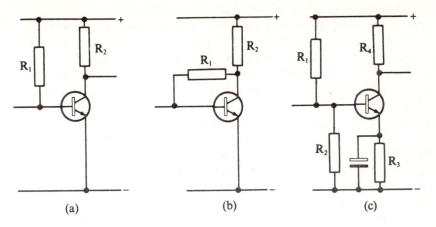

(a)   (b)   (c)

Figure 4.7   Bias systems
(a) Simple (b) Current feedback type (c) Fixed voltage type

Bias circuits

Three types of bias circuit are illustrated in Figure 4.7. The simplest uses a single resistor connected between the supply voltage and the base of the transistor (Figure 4.7(a)). This type of bias is seldom used for linear amplifier stages nowadays because it is difficult to find a suitable value of bias resistor. In this simple type of system, the value of resistor for correct biasing depends on the value of current gain (h_{fe}) of the transistor, so that a bias resistor suitable for one transistor will not work properly with another even if it is of the same type number.

Moreover, the value of bias resistor may be critical, so that one preferred value of resistance is too low and the next in the series too high.

The simple bias system is unsuitable if the transistor has to work at varying temperatures, because the voltage needed between the base and the emitter for a given collector current decreases as the temperature of a silicon transistor rises. With the simple system of bias, this change in forward voltage causes more base current to flow, and so more collector current flows as the temperature increases. Unless collector current is limited by a load resistor, the additional current will heat the transistor, so causing current flow to increase still further until the transistor is destroyed.

This process, called *thermal runaway*, is much less common nowadays, when use of silicon transistors is so general, than it was when germanium transistors found many applications.

The circuit shown in Figure 4.7(b) represents a considerable improvement, because the bias resistor is returned to the collector of the transistor rather than to the fixed-voltage supply line. This small change makes the bias to some extent self-adjusting, and the bias is now said to be *stabilized*.

The connection of the bias resistor as shown causes d.c. feedback, which means that the level of d.c. voltage at the collector affects the amount of d.c. bias current at the base. See what happens in two opposite cases. First, a change either in the transistor itself or in the load which causes collector current to increase will, because of the presence of the load resistor, cause collector voltage to drop. By Ohm's Law, this will reduce current flow through the base resistor, so reducing base current, and so reducing collector current back to near its original value. Alternatively, a change causing collector current to drop will make collector voltage rise, so passing more current through the bias resistor, causing more base current to flow, thereby increasing collector current back to nearly its original value.

All negative-feedback systems act in a similar way, tending to keep conditions in an amplifier unchanged despite other variations. A.c. feedback and its effects will be considered later on.

The third bias circuit, shown in Figure 4.7(c), is the most commonly-used of all for discrete transistor amplifier circuits. The negative feedback system of biasing is the main (and usually the only) method of biasing IC amplifiers, see later in this chapter. A pair of resistors is connected as a potential divider to set the voltage at the base terminal, and a resistor placed in series with the emitter controls emitter current flow by d.c. negative feedback. Note the emitter current is practically equal to the collector current flow ($I_c = I_e + I_b$, but I_b is very small).

In this type of circuit, the replacement of one transistor by another has little effect on the level of steady bias voltage at the collector. This biasing arrangement is therefore ideal for use in mass-produced circuits which must behave correctly even when fitted with transistors having a wide range of values of h_{fe}.

Exercise 4.2

Set up the circuit shown in Figure 4.8(a), preferably on a solderless breadboard such as the RS components range of prototyping boards.

Switch on and adjust the potentiometer until collector voltage is exactly half the supply voltage. Note this value of resistance.

Now connect a 470k resistor in parallel with the bias resistor. Note the new value of collector voltage. Remove the 470k resistor, and connect a 6k8 resistor in parallel with the load resistor. Note the value of collector voltage now. Remove the 6k8 resistor, and replace the transistor with another of the same type, again noting the collector voltage.

These three readings show how the bias voltage has been altered by changes in the bias resistor, the load resistor and the transistor respectively.

Now set up the bias system shown in Figure 4.8(b) and again adjust the potentiometer until collector voltage is exactly half the supply voltage. Note this value.

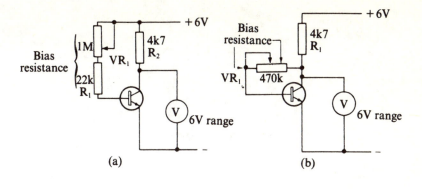

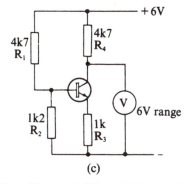

Figure 4.8 The effects of different bias systems

As before, connect a 470k resistor in parallel with the bias potentiometer and note the collector voltage. Then remove the 470k resistor, and connect a 6k8 resistor in parallel with the load resistor, noting the effect on the collector voltage. Finally, remove the 6k8 resistor, and find the effect on collector voltage of replacing the transistor. Have the changes in collector voltage been as large as they were in the first case?

Finally, set up the bias system shown in Figure 4.8(c). Measure the collector voltage, then apply the same tests as before and note the effects. Try out several types of silicon transistors in the circuit, noting the collector voltage for each. Does this circuit stabilize bias voltage well?

Figure 4.9 illustrates the circuit of a junction FET and shows the bias method which is used for a FET in the depletion mode. (MOSFETs use the same type of bias circuit also.)

For correct bias, the voltage of the gate should be negative with respect to the source voltage – or, to put the same thing another way, the source voltage must be positive with respect to the gate voltage. In this circuit the positive voltage is

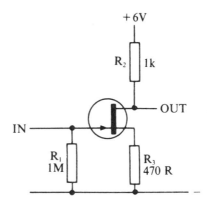

Figure 4.9   Biasing circuit for a junction FET

obtained from the voltage drop across the resistor R_3 in series with the source. The gate voltage is kept at earth (ground level, or zero volts) by the resistor R_1, which needs to have a very large value since no current flows in the gate circuit.

It will be seen that the biasing of a FET is considerably simpler than is the biasing of a bipolar transistor.

Summary

The purpose of biasing a transistor is to set its output current to a value which permits the best use to be made of its transfer characteristic.

For a linear amplifier having a resistive load, the most useful bias setting is to a collector voltage close to half the supply voltage (Class A).

The biasing method chosen must be stable, lest the bias setting be upset by small changes in component values.

Bias failure

Bias failure can be caused by either open-circuit or short-circuit bias components.

In any of the circuits shown in Figure 4.7, if a s/c develops across the resistor R_1, the large bias current that will flow in consequence will cause the collector voltage to bottom, and may burn out the base-emitter junction. If the base-emitter junction thus becomes o/c, collector voltage will rise to supply voltage, so that the same fault can be the cause of either symptom.

If R_1 becomes o/c, there is no bias supply and the collector voltage cuts off, so that collector voltage equals supply voltage.

In the case of the circuit in Figure 4.7(c), faults in resistors R_2 and R_3 can also affect the bias. The table below summarizes the possible faults and their effects.

Fault	Collector voltage	Emitter voltage
R_2 o/c	Low	High
s/c across R_2	High	Zero
R_3 o/c	High	High
s/c across R_3	Low	Zero

Gain and bandwidth

The voltage gain (G) of an amplifier is defined as:

$$G = \frac{\text{Signal voltage at output}}{\text{Signal voltage at input}}$$

with both signal voltages measured in the same way (either both r.m.s. or both peak-to-peak). This quantity G is an important measure of the efficiency of the amplifier and is often expressed in decibels by means of the equation:

$$dB = 20 \log G$$

Example

Find the gain of an amplifier in which a 30mV peak-to-peak input signal produces a 2V peak-to-peak output signal.

Solution

Insert the data in the equation: $G = \dfrac{\text{Output signal}}{\text{Input signal}}$

Then
$$G = \frac{2000}{30} = 66.7$$

Note that the 2V must be converted into 2000mV, so that both input and output signals are quoted in the same units.

Expressing the same answer in decibels:

$$G = 20 \log 66.7 = 36.5 \text{ dB.}$$

The example shows the superiority of the decibel method of expressing gain.

A decrease in gain from 66.7 to 60 might seem significant, but the same decrease expressed in decibels is only from 36.5 dB to 35.5 dB – a change of 1 dB, which is the smallest change of gain that can be detected by the ear when the amplifier is in use. Measurements of gain expressed in decibels can therefore show whether changes of gain are significant or not. Figures of voltage gain by themselves are often misleading for this purpose.

Voltage amplifiers do not have the same value of gain at all signal frequencies. Figure 4.10 shows the components in a single-stage transistor amplifier which

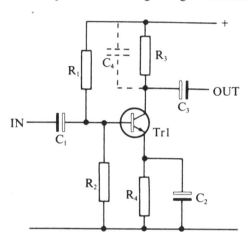

Figure 4.10   The components in a single-stage transistor amplifier which determine frequency response

determine frequency response. (C_4 is shown dotted because it consists of stray capacitances and is not an actual physical component.)

In the circuit, C_1 prevents d.c. from the signal source from affecting the bias at the base of the transistor, and C_3 prevents d.c. from its collector from affecting the next stage. The circuit can therefore give no voltage gain for d.c.; and the amount of gain it can give at low a.c. frequencies is inevitably limited by the action of the capacitors C_1 and C_3. (Gain will also be affected by C_2, because this capacitor bypasses the negative feedback action of R_4 for a.c. signals only.)

Again, at the high end of the frequency scale, the stray capacitances which would then be present at the collector of the transistor and in any circuit connected through C_3 – they are represented by C_4 connected across the load resistor – act to bypass high-frequency signals, so that gain decreases at these frequencies also. Only in the medium range of signal frequencies most commonly used is the gain given by this circuit configuration constant.

A typical curve of gain (in decibels) plotted against frequency for such an amplifier is shown in Figure 4.11. Note that the frequency scale is logarithmic, in that tenfold frequency steps occupy equal lengths of horizontal scale. This

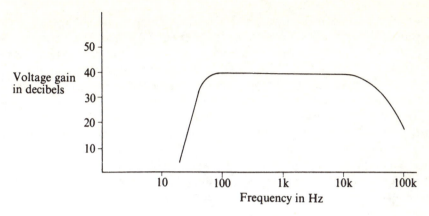

Figure 4.11   Curve of gain plotted against frequency

type of scale is necessary to show the full frequency range of an amplifier.

When a tuned circuit is used as the output load of an amplifier (Figure 4.12),

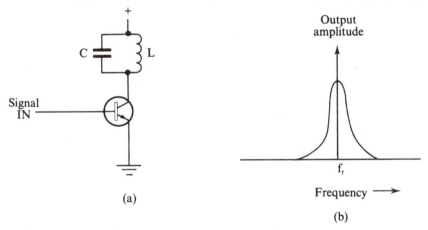

Figure 4.12   A tuned amplifier, with frequency response

the shape of the gain/frequency graph becomes more peaked. The reason is that the tuned circuit presents a high resistance to the signal at the frequency of resonance (f_r).

At resonance, this resistance has a value of L/CR ohms, where L is in henries, C in farads, and R is the resistance of the coil expressed in ohms. This value L/CR is called the *dynamic resistance* of the tuned circuit.

At all frequencies other than resonance, the load has a considerably lower value of resistance and acts instead like an impedance – with the result that voltage and current become out of phase with one another.

In such a tuned amplifier, the gain at the resonant frequency is often controlled by an *automatic gain control (AGC)* voltage. This voltage is applied at the base, either opposing or adding to the existing d.c. bias. What is known as *reverse AGC* uses a d.c. bias voltage that reduces normal bias current; while *forward AGC* uses a bias voltage that increases the normal bias current. The type of AGC used depends on the type of transistor and the circuit of which it is a component.

To give forward AGC, a resistor is included in series with the load – but bypassed by a capacitor, so that signal current does not flow through it – and the transistor is so designed that gain is much lower at low collector voltages. Increasing the bias current then lowers collector voltage and decreases the gain.

Most transistors, however, give greater gain at higher bias currents (given unchanged collector voltage), and so need to use reverse AGC.

Exercise 4.3

Construct the circuit shown in Figure 4.13, in which Tr1 can be any medium-

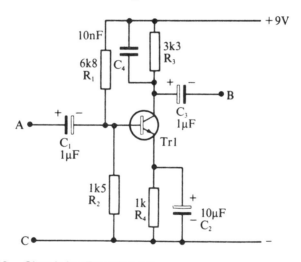

Figure 4.13   Circuit for Exercise 4.3

current NPN silicon transistor such as 2N697, 2N1711, 2N2219 or BFY50. Measure the collector voltage and note its value.

Connect the output of the signal generator to Point A, and its earth return to Point C. The Y-input of an oscilloscope should also be connected to Point A. Switch on all circuits, and adjust signal generator output to give a 1kHz, 30mV peak-to-peak signal at Point A.

Now connect the Y-input of the oscilloscope to Point B, and read the peak-to-peak value of the output signal. Calculate the gain, convert it to decibels, and record this value as the gain of 1kHz. If the output wave is noticeably distorted (i.e., flattened at either peak), reduce the amplitude of the signal generator output until a well-shaped sine-wave appears at the output. Note the new value of input signal needed.

Now multiply your known figure of output voltage at 1kHz by 0.71, and reduce the output frequency of the signal generator until the output voltage reaches this lower value. Record the signal generator frequency required to achieve this, and call it f_1.

Increase signal generator frequency again until you find a frequency above 1kHz at which output voltage is reduced by the same amount. Record this frequency as f_2. Note that during both these readings, the output of the signal generator should remain constant at its original value of amplitude. Check the output amplitude at both frequencies, f_1 and f_2, if there is any doubt.

The factor of 0.71 which you applied above corresponds to a loss of gain of 3dB. The frequencies f_1 and f_2 are therefore called the *lower* and the *upper 3dB points* respectively, and the range of frequencies between them is called the *bandwidth* of the amplifier. For an audio amplifier, f_1 and f_2 are quoted, so that such an amplifier can be described as being (for example) '3dB down at 17Hz and at 35kHz'.

In short, the term 'bandwidth', as applied to a tuned amplifier, means the quantity $f_2 - f_1$. A tuned amplifier can therefore be described as having (again for example) 'a bandwidth of 10kHz centred on f_r at 470kHz'.

The purpose of the 10nf capacitor C_4 connected across the load resistor R_3, is to ensure that the response of the amplifier at high frequencies will not be too wide.

Exercise 4.4

Still using the circuit of Figure 4.13, observe the changing output waveform while signal generator output at 1kHz is increased to 300mV. Sketch the waveform which shows the distortion caused by overloading.

Next, restore the input to its previous value and connect an 8k2 resistor across R_1. Sketch the resulting output waveform, which will show the distortion caused by over-biasing.

With the resistor across R_1 removed, connect a 680-ohm resistor across R_2. Sketch the output waveform, which now shows the distortion caused by under-biasing. Note the changes in voltage readings when the following conditions are present or simulated (bracketed actions):

(a) base-emitter of Tr1 o/c (disconnect lead to base)
(b) base-emitter of Tr1 s/c (short emitter to base connections)
(c) C_2 s/c (short across connections).

The forms of distortion caused by overloading or by faulty biasing will be obvious when the output waveforms are viewed. Note, however, that smaller amounts of distortion caused by curvature of the characteristics are not visible on an output trace, and can only be detected by *distortion meters*. These filter out the sine-wave which is being amplified, leaving an output which consists only of the distortion, which can then be measured.

The table below lists faults which have predictable effects, especially on **gain** and bandwidth.

Fault	Effects
Emitter bypass capacitor o/c	Reduced gain: increased bandwidth
Collector load resistance too low	Low gain: collector voltage abnormally high
Transistor under-biased	Gain reduced
	Some signal distortion at output

Exercise 4.5

Construct the junction FET amplifier circuit shown in Figure 4.14. The variable

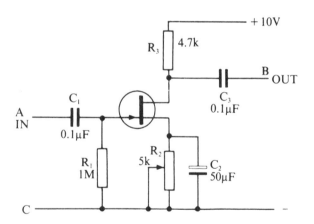

Figure 4.14   Circuit for Exercise 4.5

resistor R_2 should be set so that drain voltage is about 7V when there is no signal output.

Using the methods indicated in Exercise 4.3, find the gain at 1kHz, the frequencies which produce the -3dB points, and note the distortion caused by overloading. Note the changes in voltage readings and (if applicable) gain when the following conditions are present or simulated by removing or short-circuiting components:

(a) FET faulty (o/c channel or s/c gate to source or drain)
(b) C_2 o/c
(c) C_2 s/c
(d) R_2 high (more than 47k)
(e) R_3 high (more than 470k).

Summary

$$\text{Voltage gain} = \frac{\text{Signal voltage out}}{\text{Signal voltage in}}$$

In decibels, this becomes

$$\text{Voltage gain} = 20 \log \frac{\text{Signal voltage out}}{\text{Signal voltage in}}$$

The graph of voltage gain in decibels plotted against frequency shows the -3dB points at which the bandwidth of the amplifier is measured.

The bandwidth is taken as the useful operating frequency range of the amplifier.

Multiple-stage amplification

In many applications, a single transistor is not enough to provide sufficient gain, and several stages of amplification are needed. When an amplifier contains several stages, its total gain is given by the equation:

$$G = G_1 \times G_2 \times G_3$$

where G_1, G_2, G_3 are the gains of the individual stages.

In decibels, this becomes

$$(dB)_1 + (dB)_2 + (dB)_3$$

the total gain in decibels. Note that the decibel figures of gain are *added*, whereas the voltage (or current, or power) figures have to be *multiplied*. This is because logarithms are used in the construction of the decibel figure, and logarithmic addition is the equivalent of the multiplication of ordinary figures.

The coupling together of separate amplifying stages involves transferring the output signal from one stage to the input of the next stage. This can be done in several ways, as described below and illustrated in Figure 4.15 (from which details of all biasing arrangements have been omitted for the sake of clarity):

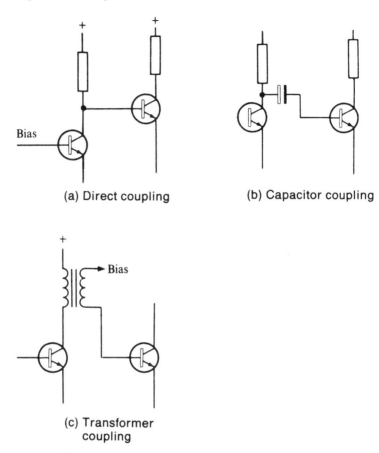

(a) Direct coupling

(b) Capacitor coupling

(c) Transformer coupling

Figure 4.15   Coupling amplifying stages

(a) *Direct coupling* involves connecting the output of one transistor to the input of the next, using only resistors or other components which will pass d.c. The result is that both d.c. and a.c. signals will be coupled. A d.c.-coupled amplifier by definition amplifies d.c. signals, so that a small change in the steady base voltage of the first stage will cause a large change in the steady collector voltage of the next. In all d.c.-coupled stages, particular attention needs to be paid to bias. A negative feedback biasing system is usually required.

(b) *Capacitor coupling* makes use of a capacitor placed in series between the output terminal of one stage and the input terminal of the next. The effect is that a.c. signals only can be coupled in this way, because d.c. levels cannot be transmitted through a capacitor. When amplifiers need a low -3dB point when input frequency is only a few Hz, large values of capacitance will be required.

(c) *Transformer coupling* makes use of current signals flowing in the primary winding of a transformer connected into the collector circuit of a transistor to induce voltage signals in the secondary winding, which in turn is connected to the base of the next transistor. Once again, only a.c. signals can be so coupled; and a well-designed transformer will be needed if signals of only a few Hz are to be coupled. Note that the gain/frequency graph of a transformer-coupled amplifier can show unexpected peaks or dips caused by resonances. For that reason, transformer-coupled amplifiers are seldom used when an even response is of importance.

The use of negative feedback

Though it is possible to design single amplifier stages with fairly exact values of voltage gain (it is, for example, quite possible to design an amplifier with a voltage gain of exactly 29 times, if that should happen to be wanted), it is less easy to design multi-stage amplifiers that will give the precise voltage gain required. The reason is that the input and output resistances of transistors vary considerably, depending on the varying h_{fe} values of individual transistors, and that in any form of signal coupling the output resistance of one transistor forms a voltage divider with the input resistance of the next – so attenuating the signal. The point is simply illustrated in Figure 4.16, in which R_1 symbolizes the output

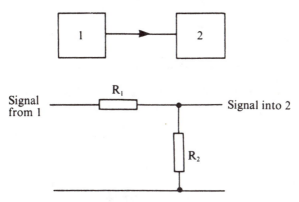

Figure 4.16   Potential divider network

resistance of the first transistor and R_2 the input resistance of the second.

A more promising way of designing an amplifier for a specified figure of gain is to aim for one which has too large a value of voltage gain, and then to use negative feedback to reduce this gain to the required figure. One advantage of using negative feedback is that it is often possible to calculate the gain of the complete amplifier without knowing any of the individual transistor gains or resistances. Very often the gain of the amplifier is simply the ratio of two values of fixed resistors. Some feedback methods are shown in Figure 4.17.

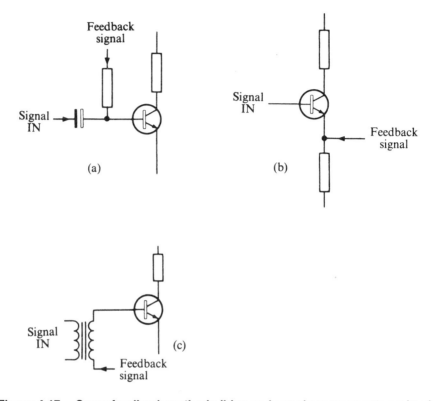

Figure 4.17   Some feedback methods (bias and supply components omitted to aid clarity)
(a) Shunt feedback into the base (b) Series feedback into the emitter (c) Series feedback using a transformer

For negative feedback to be useful, the gain of the amplifier without feedback (called the *open-loop gain*) must be much greater (100 times or more) than the gain of the amplifier when feedback is applied (the so-called *closed-loop gain*). When this situation exists, the factor $\dfrac{\text{Closed-loop gain}}{\text{Open-loop gain}}$ becomes important, because distortion and noise generated in the transistors or other components within the amplifier will be reduced by exactly the same ratio. If, for example, open-loop gain is 10 000, and closed-loop gain 100, then the factor $\dfrac{\text{Closed-loop gain}}{\text{Open-loop gain}}$ becomes $\dfrac{10}{10\,000}$, or $\dfrac{1}{100}$ which means that the distortion will be reduced to 1/100th of its value in the open-loop amplifier.

Bandwidth, on the other hand, will be increased by the factor
$$\frac{\text{Open-loop gain}}{\text{Closed-loop gain}}$$

94

(which is 100 times in the example above), on the assumption that no outside factor exists to limit bandwidth in any other way.

There are four principal ways in which negative feedback can be applied to an amplifier, depending on where the signal is taken from and to what point it is fed back.

1 A signal fed back from a collector is called *voltage-derived*, because it is a sample of the output voltage signal.
2 A signal fed back from a resistor in the emitter circuit is called *current-derived*, because it is proportional to the output current signal and in the same phase as it.
3 A signal fed back to a base input is called a *shunt feedback signal*, because the feedback signal is in shunt (or parallel) with the normal input signal.
4 A signal fed back to the emitter of an input transistor is called a *series feedback signal*, because the feedback signal is in series with the normal input signal. An alternative method of achieving a series feedback signal is to transformer-couple the feedback to the base of a transistor through the transformer winding, as shown in Figure 4.17(c).

Each of these methods of applying negative feedback will have the desired effects of reducing gain, noise and distortion, and of increasing bandwidth; but the different methods of connection can affect other features of the complete amplifier. Taking the feedback from the emitter of an output stage, for example, causes the output resistance at the collector of the same transistor to be higher than it would be if the feedback were to be taken from the collector. Alternatively, taking the feedback to a base input causes the input resistance to be lower (often much lower) than it would be if the feedback were to be taken to the emitter.

Some feedback circuits include the input or output resistance of the transistor itself as part of the feedback loop, and are therefore less predictable in action.

Figure 4.18 shows two common types of feedback circuit. Figure 4.18(a) uses negative feedback from the emitter of Tr2 to the base of Tr1. The feedback is therefore series-derived and shunt-fed.

Figure 4.18(b) shows negative feedback from the collector of Tr2 to the emitter of Tr1. The feedback is therefore shunt-derived and series-fed.

The table below shows the effect of some component failures on both these circuits.

Fault	Effects
R_9 or C_5 o/c	No negative feedback, high gain but possible instability
s/c across R_9	Greatly reduced gain
C_5 s/c	Bias of Tr1 incorrect

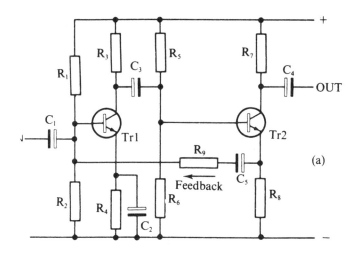

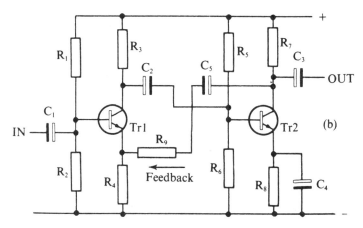

Figure 4.18   Two common types of feedback circuit

Exercise 4.6

Construct the circuit shown in Figure 4.19. Both Tr1 and Tr2 can be any general-purpose silicon NPN transistor. Then observe the effects of two different types of feedback.

(a) *Series-derived, shunt-fed.* Measure the voltage gain of the complete amplifier. (Note that a small input signal should be used to prevent overloading). Now remove C_4, so introducing negative feedback through R_1. Measure and note the new value of voltage gain.

(b) *Shunt-derived, series-fed.* Replace C_4, remove C_2, and again measure the

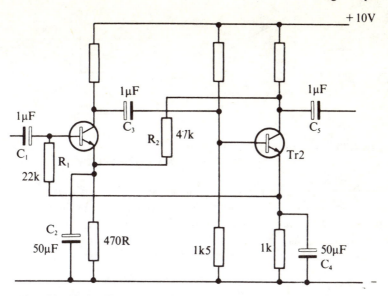

Figure 4.19   Circuit for Exercise 4.6

value of circuit gain. This gain is now the loop gain of the feedback loop through R_2. Note its value and compare with (a) above.

Summary

If more gain is required than can be achieved by a single-stage transistor amplifier, several stages of amplification can be coupled together.

Coupling can either be direct, or achieved by means of capacitors or transformers.

The value of gain can be fixed precisely by using negative feedback. Negative feedback reduces gain, but also reduces the effect of component variations on that gain. Correctly applied, negative feedback also reduces distortion and noise, and widens bandwidth.

Differential amplifiers

A circuit much used in industrial equipment is the *balanced* or *differential, amplifier*.

A balanced amplifier has two inputs, one the phase-inverted 'image' of the other. The output often (but not always) appears in similar balanced form. Figure 4.20 shows an amplifier with balanced output (a), and with unbalanced output (b). Balanced output signals are said to be 'balanced about earth' if the sum of the instantaneous voltages is always zero.

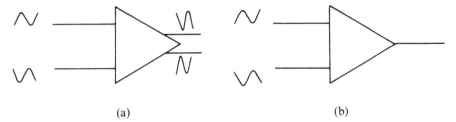

(a)   (b)

Figure 4.20   The differential amplifier
(a) Balanced and (b) Unbalanced outputs

The value of balanced signal amplifiers lies in the fact that such amplifiers make it possible to amplify small (balanced) signals even in the presence of very large (unbalanced) noise pulses, so that the amplifier output contains only the wanted signal. Thus ideally there should be no output signal for any common mode input signal. The comparison of these two inputs is shown in Figure 4.21.

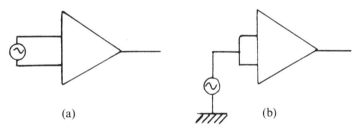

(a)   (b)

Figure 4.21   Differential and common mode input

The ratio of amplifier voltage gain for each type of signal is called the *common mode rejection ratio*, and is generally expressed in decibels by means of the following formula:

$$\text{Common mode rejection ratio} = 20 \log \frac{G_b}{G_{cm}}$$

where G_b = voltage gain of amplifier for balanced signals, and
G_{cm} = voltage gain of amplifier for unbalanced signals.

Alternatively stated, the CMRR is the ratio of the input signal levels needed to produce the same level of output signal in each case.

In this context, the phrase 'common-mode' signifies an unbalanced signal appearing at both input terminals in the same phase.

Values of common-mode rejection ratio of 100 dB and more can be obtained by suitable amplifier design, which means that voltage gain for the balanced signal is 100 000 times, or more, the voltage gain for the unbalanced signal. Such large values of common-mode rejection ratio render industrial amplifiers virtually insensitive to interference, provided only that the wanted signal is in balanced form.

This wanted signal is generally derived from some sort of transducer, and many types of transducers can be connected so that they give balanced output signals. The various types of bridge connection, for example, can all be arranged to produce a balanced signal; and inductive transducers can be wound with a centre-tap so that their outputs are always balanced.

The basic balanced amplifier circuit, known either as the *differential amplifier* or as the *long-tailed pair*, is shown in Figure 4.22 (with the biasing circuits

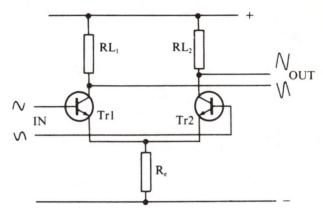

Figure 4.22   The basic long-tailed pair circuit

omitted for the sake of clarity). The resistor R_e is an essential part of the circuit because, when the circuit is correctly biased and operated, the signal voltage across it is always zero. An increase in (instantaneous) voltage at the base of one of the transistors ought to be accompanied by an identical decrease at the base of the other. A signal will therefore appear across R_e only when the inputs are not perfectly in antiphase.

R_e therefore forms part of the action which discriminates against common-mode signals. It is the 'tail' of the 'long-tailed pair' and intended to pass a constant current which is shared between Tr1 and Tr2.

In a balanced amplifier, the signal voltage is applied equally between the two base inputs, and the amplified output signal is taken from between the collectors. Given that the transistors have identical values of mutual conductance (they form a matched pair), any signal which appears identically at both bases will give rise to an amplified and inverted signal at both collectors; but the voltage between the collectors will be unaltered (see Figure 4.23). It is this action which provides the rejection of the noise signals – a large enough value of R_e ensuring that any common-mode signal arising from a lack of balance between the transistors is offset by negative feedback from the signal voltage across R_e.

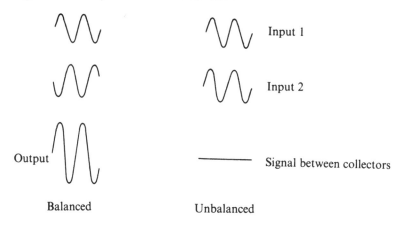

Figure 4.23   Signals associated with long-tailed pair amplifiers

Hum or other interference on the supply line is treated as the common-mode signal which in fact it is, and so produces no output.

The output of a differential amplifier can be used to drive another similar stage, or it can be converted to unbalanced form by using one of the circuits shown in Figure 4.24 (bias circuitry again omitted).

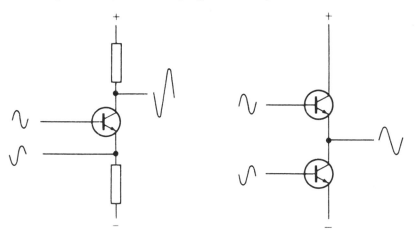

Figure 4.24   Converting balanced inputs to an unbalanced output

The differential amplifier is also a versatile circuit for the handling of unbalanced input signals. If one transistor base in the circuit is connected (by a capacitor, for example) to signal earth, an unbalanced signal applied to the base of the other transistor will result in a balanced signal between the two collectors (Figure 4.25). The voltage gain is approximately equal to that of a single transistor operating at the same bias current. A single-ended output can also be

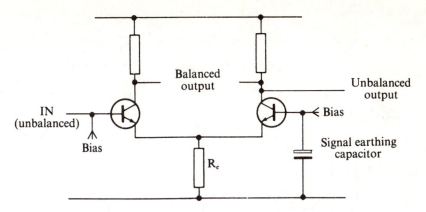

Figure 4.25   Unbalanced input: output either balanced or unbalanced

obtained, but the voltage gain will then be restricted to half what it would have been if the output has been a balanced one from the same circuit. Input resistance is high in both these circuits, and the emitter resistor acts as the coupling between the two transistors.

The Figure 4.25 circuit is particularly useful in practice, for even with an unbalanced input the circuit discriminates against interference pulses and supply hum affecting both bases alike.

Note that in the Figure 4.25 circuit, the output signal is in phase with the input signal. A significant improvement in performance can be obtained if the resistor R_e is replaced by another transistor, so biased as to keep the current flowing into the differential pair constant. This circuit variation (shown in Figure 4.26 with the biasing components again omitted) is much used.

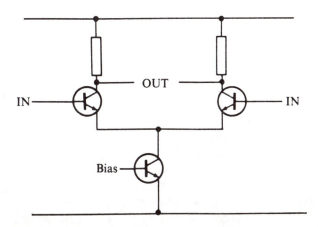

Figure 4.26   A transistor connected as 'the tail' improves performance

Exercise 4.7

Assemble the long-tailed pair circuit shown in Figure 4.27 using the following

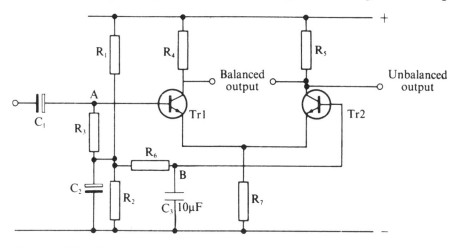

Figure 4.27   Circuit for Exercise 4.7

component values: $R_1 = 12k$; $R_2 = 2k2$; R_3 and $R_6 = 56k$; R_4 and $R_5 = 6k8$; $R_7 = 1k$; C_1, C_2 and $C_3 = 10\mu F$, 25 v.w. Tr1 and Tr2 can be any small-signal silicon NPN voltage amplifier transistors.

Connect the circuit shown to a power supply in which neither positive nor negative line is earthed. Measure the collector and emitter voltages relative to the negative line, and note the values. Connect the inputs to an unearthed signal generator output and the differential outputs to an oscilloscope. Note that if either the power supply or the signal generator has one lead earthed, the signals will be shorted out.

Electrical safety requirements are most easily met if both the circuit and the signal generator are battery-operated. If an oscilloscope with a differential input is available, an earthed signal generator can be used provided that a battery supply is used for the differential amplifier.

Measure the gain at 400Hz, and note this value as the value of differential gain, G_b.

Now break the circuit at Point B, and reconnect the base of Tr2 to Point A. Again measure the gain at 400Hz. The result will give the common-mode gain, G_{cm}. Calculate the common-mode rejection ratio in decibels.

Re-connect the oscilloscope and measure the output signal between the collector of Tr2 and the negative line. A single-ended output is now being taken. Re-connect Tr2 base to Point B, find and note the gain at 400Hz. Now break the circuit again at Point B and connect the base of Tr2 to Point A once more. Find the value of gain for this arrangement.

Is there any significant common-mode rejection when a single-ended output is used?

Operational amplifiers

Operational amplifiers (commonly known as *op-amps*) were originally developed to perform mathematical operations in analogue computers. Until the advent of integrated circuits however, their use in discrete form was very much limited to these functions alone. Due to mass production techniques employed for integrated circuits, op-amps were found to have many very useful properties for other applications. Consequently op-amps are now found in a very wide variety of equipment.

IC amplifiers are multi-stage units with direct coupling and very large values of gain. The *open-loop gain* is the amount of gain that would be obtained by using the IC as an amplifier without feedback, but this amount is so large (usually 100 dB or more, corresponding to voltage gain of 100 000 or more) as to be unusable. IC amplifiers are always used with feedback, and since direct coupling is normally used, the feedback is used both to establish bias and to establish signal gain.

Advantages of using integrated circuits

1   Increased reliability due to a construction that leads to fewer interconnections to develop 'dry joints'.
2   Reduced size compared with their discrete component counterparts.
3   Reduced costs with volume production.
4   Faster operation or better high-frequency response due to shorter signal path lengths.
5   IC fabrication gives a better control over the spread or variation of device parameters.
6   Reduced assembly time due to fewer soldered joints.
7   Since the final performance of the circuit can be closely controlled by using negative feedback, the overall design of circuits is simpler.

A typical example is the so-called *Type 741*. The remainder of this section will describe the operation of the IC version of this circuit.

The ideal op-amp would have the following characteristics:
• very high input resistance;
• very low output resistance;
• very high voltage gain, open-loop; and
• very wide bandwidth.

Figure 4.28 shows typical characteristics for a 741 op-amp. It will be seen that

Input resistance	1M
Output resistance	150R
Open-loop gain	100dB
Bandwidth	1kHz at a gain of 60dB
Maximum supply voltage	± 18V (dual supply)
Maximum load current	10mA

Figure 4.28 Typical characteristics of an operational amplifier type 741

the first three of the above requirements are easily met; while the bandwidth, though small by audio standards, is reasonably wide having regard to the signals which the 741 was designed to handle – namely, signals ranging from d.c. to a few hundred Hz.

The gain-frequency response of a typical op-amp is shown in Figure 4.29.

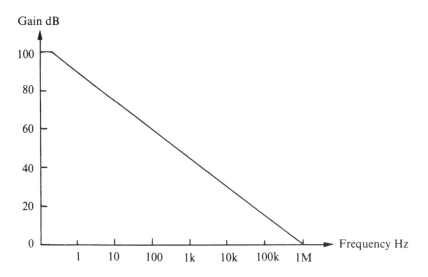

Figure 4.29 Gain-frequency response of an op-amp

Remember that by using negative feedback the stage gain of an amplifier is reduced but at the same time its bandwidth is increased. This is due to the fact that the amplifier's gain bandwidth product is a constant. This feature of being able to trade gain for bandwidth is probably the most important feature of the op-amp and forms one of the design steps in the production of a practical amplifier.

The circuit symbol for an op-amp is shown in Figure 4.30. Two power

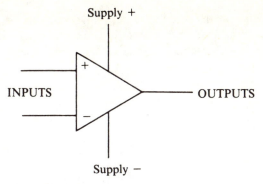

Figure 4.30   Circuit symbol for an op-amp

supplies are needed (though single supply lines can also be used by modifying the circuits). The dual supply is balanced about earth. The op-amp has two inputs, labelled ($+$) and ($-$) respectively, and a single output. The internal circuit is that of a balanced amplifier, so that the output voltage is an amplified copy of the voltage difference between the two inputs. The ($+$) sign at one input indicates that feedback from the output to this input will be in phase. The ($-$) sign similarly indicates that feedback to this input will be out-of-phase.

The voltage gain, open-loop, is some 10^5 (100 dB) or more. Since the internal circuit is completely d.c. coupled, both d.c. and a.c. negative feedback will be needed for linear amplifier applications.

Op-amp circuits

Figure 4.31 illustrates a basic phase-inverting amplifier circuit. Balanced power supplies are used, the ($+$) input being earthed and the ($-$) input connected to the output by a resistor R_f which provides negative feedback of both d.c. and signal voltages. R_{in} serves to increase the input resistance.

Although the ($-$) input is not actually connected to earth, its voltage (either d.c. or signal) is earth voltage, and it is said for this reason to be a 'virtual earth'. What happens is this. The ($+$) input is earthed, and any voltage difference between the inputs is amplified. Suppose, for example, that the ($-$) input is at 0.1mV below earth voltage. This 0.1mV will be amplified by 10^5, giving a $+$ 10V output (note the phase change). This voltage then drives current through R_f to increase the voltage at the ($-$) input until it equals the voltage at the ($+$) input.

Similarly, a 0.1mV positive voltage at the ($-$) input would cause the output to swing to $-$ 10V, again causing the input to return to zero.

Two points should be noted:

1   The voltage gain is so high that the assumption that the ($-$) input is at zero voltage is justified.

2 Any unbalance in the internal circuit will cause an 'offset' voltage at the input, so that the (−) input may need to be at a small positive or negative voltage to maintain the output at zero (the input offset voltage).

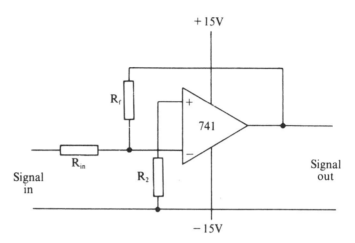

Figure 4.31   A basic phase-inverting amplifier

This input offset is reduced in the Type 741 by adding the *offset balancing resistor* shown in the circuit of Figure 4.32 (in which the small figures are the IC pin numbers). The size of the offset voltage will, however, vary as the temperature of the IC varies, so that some compensation may be needed in circuits such as high-gain d.c. amplifiers, in which offset can be troublesome. This is known as the input offset voltage *temperature drift* usually abbreviated to *drift*.

Due to the virtual earth at the (−) input of the basic circuit in Figure 4.31 the circuit input resistance becomes equal to R_{in}. It can also be shown that the stage gain is simply the ratio of R_f/R_{in}. To minimize the effects of temperature drift, the resistor R_2 is made equal to the parallel combination of R_f and R_{in}, that is, $\frac{R_f \times R_{in}}{R_f + R_{in}}$. Again due to the virtual earth, several input signals can be applied to the (−) input through resistors. This forms the basis of the *summing amplifier*, where the input signals are added.

Figure 4.33 shows a type of circuit known as a *follower*. In this circuit the signal input is applied to the (+) input, whose d.c. value is normally fixed at earth voltage by R_1. The feedback resistor R_2 ensures that the (−) input is a virtual earth. A signal coming into the input will now produce an identical signal, in phase, at the output. The input resistance of the circuit is extremely high, the output resistance very low.

Unlike the familiar emitter-follower, this type of circuit can be modified to

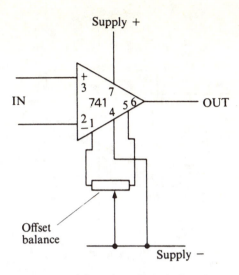

Figure 4.32   The op-amp: adding the offset balancing resistor

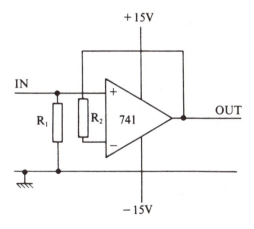

Figure 4.33   A follower circuit: no voltage gain

produce voltage gain, as is shown in Figure 4.34. Such a circuit should only be used when the signal input is of small amplitude because the effect of the feedback is to make the input signal a common-mode signal, and the amplitude of common-mode signals needs to be kept below a specified value in most op-amp designs.

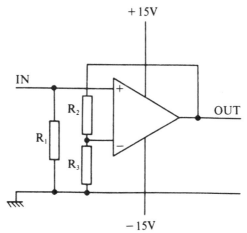

Figure 4.34   A follower circuit: voltage gain $= \frac{R_2}{R_3} + 1$

The op-amp integrator

The characteristics of the op-amp make it possible to build integrating circuits giving much better performance than did integrators using only passive components.

Figure 4.35 shows a simple op-amp integrator which functions also as a low-

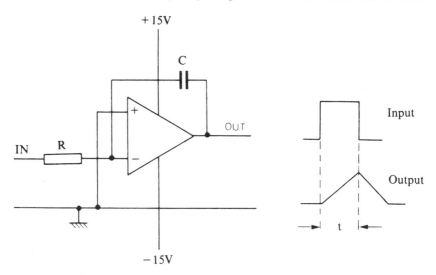

Figure 4.35   The op-amp integrator: basic circuit

pass filter. Provided that the time constant CR seconds is approximately five times the periodic time 't' of the input signal, then the output waveform will have the shape shown in Figure 4.35.

A fully practical circuit, however, must also include some method of setting the output voltage to zero before the circuit is used, particularly if the d.c. level of the output is important. If the integrator is to be used for a.c. only, a bias resistor (as R_1 in Figure 4.36) will serve to prevent drift.

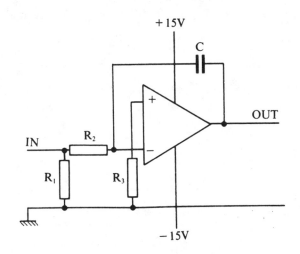

Figure 4.36   The op-amp integrator: practical circuit

The op-amp differentiator

If C and R of Figure 4.35 are interchanged and their time constant is approximately one fifth of the periodic time 't', the circuit becomes a differentiator. The output waveform then depends on the rate of change of the input signal as

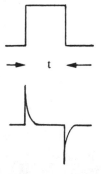

Figure 4.37   Input–output waveforms of a differentiator

shown in Figure 4.37. For input signals of a different shape, the differentiator acts as a high-pass filter.

The differential amplifier

When an op-amp is operated in the differential mode as in Figure 4.38, its

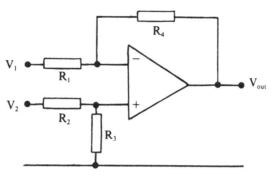

Figure 4.38   The differential amplifier

output V_{out} is proportional to the difference between the two inputs V_1 and V_2. The polarity of V_{out} depends upon which input is the larger. If V_1 is the greater, then V_{out} is negative.

Applications of op-amps

Op-amps can thus be used in many roles, the following being typical:

1   As differential amplifiers, having exceptionally good common-mode rejection ratios, and with response down to d.c.
2   As amplifiers for audio frequencies whose gain is controlled entirely by the values of the feedback components.
3   As integrators and differentiators for the shaping of signal waveforms.
4   As sensitive switching circuits, triggered by the very small size of the differential input voltage needed to switch the output from one level to another.
5   As comparators, providing an output which is an amplified version of the voltage difference between two signal inputs (a.c. or d.c.).

Exercise 4.8

Construct the two op-amp circuits shown in Figure 4.39 (which are both shown, for convenience, as operating from a single-ended power supply).

   In each circuit, measure the d.c. voltages with a high-resistance meter, both at

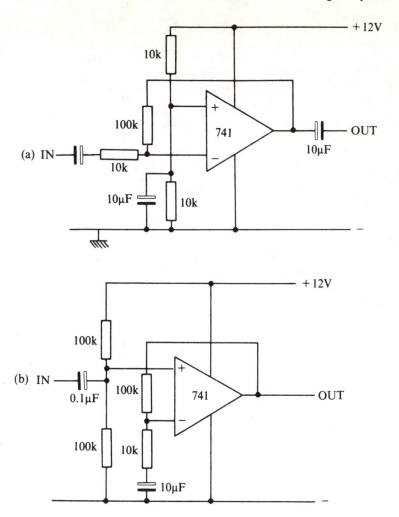

Figure 4.39   Circuits for Exercise 4.8

the output of the 741 and at each input. Then, with the aid of a signal generator and an oscilloscope, measure the amount of gain at 400Hz.

Does the amount of gain shown correspond to what you would expect to get from the values of the resistors in the two circuits?

Exercise 4.9

Connect up the circuit in Figure 4.40 which uses the 741 op-amp as a differential amplifier. The input arrangement makes it possible to apply signals of differing amplitudes to the two inputs, and the use of feedback resistor R_4 maintains the

111

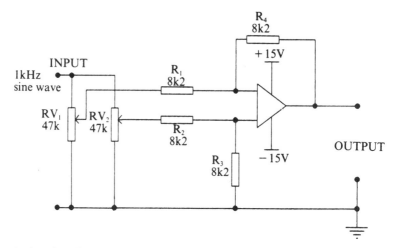

Figure 4.40   A differential amplifier circuit which uses two potentiometers to create a signal difference at the input

gain at a low value.

(a) Using the values as shown in the diagram, set the signal-generator and the potentiometers so that an output of about 2V p-p is obtained. Measure the peak-to-peak signal voltage levels at the two inputs to the 741, pins 2 and 3. Write your measurements in the form of a table and calculate the *differential gain*.

(b) Replace R_2 and R_4 by 820k resistors and repeat your measurements. What value of gain does this change cause?

Exercise 4.10

Connect up the op-amp integrator circuit as shown in Figure 4.41. Arrange for a square-wave input of 5V peak-to-peak at 500Hz from the signal generator.

(a) Using the circuit as shown, measure the peak-to-peak output voltage, and sketch the output waveform.

(b) Change the value of C_2 to 220nF and repeat your measurements.

Multiple-choice test questions

1   Increasing the value of resistive load on a single-stage voltage amplifier will:
(a) increase the bias current
(b) increase the gain
(c) decrease the gain
(d) increase the bandwidth.

2   Feedback of signal from the collector of a transistor to the base of the same

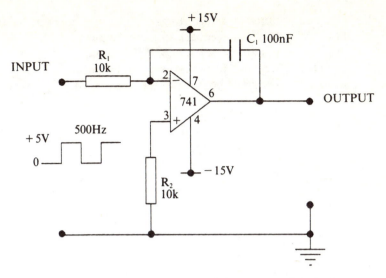

Figure 4.41   A circuit which uses the 741 opamp as an integrator

transistor will cause:
(a) oscillation
(b) more distortion
(c) reduced gain
(d) reduced bandwidth.

3   If a transistor amplifier is to be used at d.c. and very low frequencies, then it cannot make use of:
(a) negative feedback
(b) differential amplification
(c) balanced power supplies
(d) capacitor coupling.

4   An ideal operational amplifier would have:
(a) very high output impedance
(b) very high input impedance
(c) very low gain
(d) very low common mode rejection ratio.

5   At a virtual earth point there will be:
(a) negligible d.c. voltage
(b) negligible feedback
(c) negligible a.c. voltage
(d) negligible offset.

6   When an op-amp is used as an inverting amplifier, it is connected with:
    (a) input to the + terminal, feedback to the − terminal
    (b) input to the − terminal, feedback to the + terminal
    (c) input to the + terminal, feedback to the + terminal
    (d) input to the − terminal, feedback to the − terminal.

5 Power amplifiers and power supply units

Summary

Power amplification, heat sinking, classes of amplification, single-ended and push–pull (balanced) stages. Complementary stages, single-ended push–pull. Impedence matching. IC power amplifiers. Power supply regulation. Regulator circuits. Zener diode use. Series regulator. Voltage multipliers. Thyristor-switched supplies. Switched-mode supplies. Protection circuits. Line interference suppression. Battery charging.

Both voltage amplifiers and current amplifiers play important parts in electronic circuits, but not all of them are capable of supplying some types of load. For instance, a voltage amplifier with a gain of 100 times may be well able to feed a 10V signal into a 10k load resistance, but quite incapable of feeding even a 1V signal into a 10-ohm load. A current amplifier with a gain of 1000 may be excellent for amplifying a 1µA signal into a 1mA signal, but cannot convert a 1mA signal at 10V into a 1A signal at the same voltage.

The missing factor common to both these examples is power. A signal of 1A (r.m.s.) at 10V (r.m.s.) represents a power output of 10W, and the small transistors which are used for voltage or current amplification cannot handle such levels of power without overheating.

Transistors intended to pass large currents at voltage levels of more than a volt or so must have the following characteristics:

1 They must have low output resistance.
2 They must have good ability to dissipate heat.

115

A low output resistance is necessary because a transistor with a high output resistance will dissipate too much power when large currents flow through it. Low resistance is achieved by making the area of the junctions much larger than is normal for a small-signal transistor.

The ability to dissipate heat is necessary in order that the electrical energy which is converted into heat in the transistor can be easily removed. If it were not, the temperature of the transistor (notably its collector-base junction) would keep rising until the junctions were permanently destroyed.

Given suitable transistors with large junction areas and good heat conductivity to the metal case, the problem of power amplification becomes one of using suitable circuits and of dissipating the heat from the transistor.

In practice, most power amplifier stages are required to provide mainly current gain, since the voltage gain can be obtained from low-current stages before the power-amplifier stage. The power amplifier stage can therefore have very low voltage gain or even attenuate the voltage level.

Heat sinks

Heat sinks take the form of finned metal clips, blocks or sheets which act as convectors passing heat from the body of a transistor into the air. Good contact between the body of the transistor and the heat sink is essential, and *silicone grease* (also called *heat-sink grease*) is useful in promoting this contact.

Many types of power transistor have their metal cases connected to the collector terminal. It is therefore necessary to insulate them from their heat sinks. This is done by using thin mica washers between transistor and heat sink, with insulating bushes inserted on the fixing-bolts in addition. Heat-sink grease should always be smeared on both sides of all such washers. Heat sinks are illustrated in Figure 5.1.

Class A and Class B

Several different methods exist for biasing transistors which are to be used in power output stages. 'Class A' and 'Class B' are names given to two types of commonly used biasing systems.

In a Class A stage, the transistor is so biased that the collector voltage is never bottomed, nor is current flow cut off. Output current flows for the whole of the input cycle. It is the same bias system as is used in linear voltage amplifiers.

Class A operation of a transistor ensures good linearity, but suffers from two disadvantages:

- A large current flows through the transistor at all times, so that the transis-

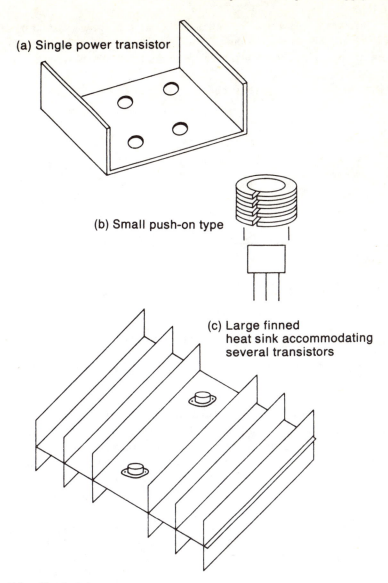

(a) Single power transistor

(b) Small push-on type

(c) Large finned
heat sink accommodating
several transistors

Figure 5.1   Heat sinks

tor needs to dissipate a considerable amount of power.
• This loss of power in the transistor inevitably means that less power is available for dissipation in the load, and a Class A stage can never be more than about 30% efficient.

Even in an ideal Class A amplifier, with the load and amplifier output

117

resistances matched, only some 50% of the available power would be delivered to the load, with the remaining 50% being dissipated by the amplifier in the form of heat. When the resistances are mismatched, the power transfer ratio is even lower.

In a Class B amplifier (see Figure 5.2), the power transistor conducts for only

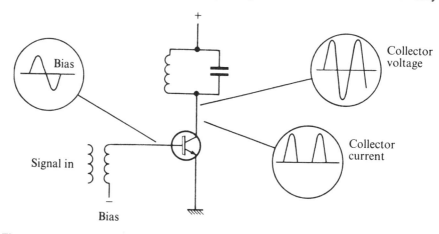

Figure 5.2   A Class B radio-frequency amplifier

one-half of the duration of the input sine-wave. A single transistor is therefore unusable unless the other half of the wave can be obtained in some other way.

At radio frequencies this can be done by making use of a load which is a resonant circuit (a so-called *tank-circuit*). The tank circuit is made to oscillate by the conduction of the transistor, and the action of the resonant circuit continues the oscillation during the period when the transistor is cut off.

This principle can only be applied, however, if the signal to be amplified is of fairly high frequency. This is because the values of inductance and capacitance needed to produce resonance at, say, audio frequencies would be too large to be practical; and because a load of such a size would in any case too greatly restrict the bandwidth.

An alternative method which is used at audio and other low frequencies is to use two transistors, each conducting on different halves of the input wave. Such an arrangement is called a *push–pull* circuit. Push-pull circuits can be used in Class A amplification, but are essential for use in Class B. (See Figure 5.3.)

Class B stages possess the following advantages over Class A:

1   Very little steady bias current flows in them, so that the amplifier has only a negligible amount of power to dissipate when no signal is applied.
2   Their theoretical maximum efficiency is 78.4%, and practical amplifiers can achieve efficiencies of between 50% and 60%. This means that more power is dissipated in the load itself, and less wasted as heat by the transistors.

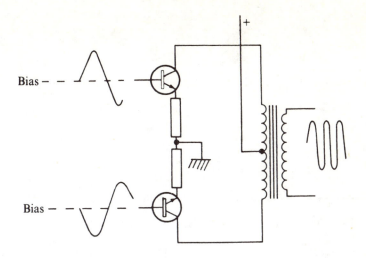

Figure 5.3   A push–pull circuit

The disadvantages of Class B compared to Class A are that:

1   The supply current changes as the signal amplitude changes, so that a stabilized supply (see below) is often needed.
2   More signal distortion is caused, especially at that part of the signal where one transistor cuts off and the other starts to conduct. This part of the signal is called the *cross-over region.*

Typical Class A and Class B circuits

Figure 5.4 illustrates a single-transistor amplifier biased in Class A. Because no d.c. can be allowed to flow in the load (which is in this case a loudspeaker), a transformer is used to couple the signal from the transistor to the load.

With no signal input, the steady current through the transistor is about 50mA and the supply voltage is 12V. Because of the low resistance of the primary winding of the transformer, the voltage at the collector of the transistor is about equal to the supply voltage.

When a signal is applied, the collector voltage swings below supply voltage on one peak of the signal and the same distance above the supply voltage on the other peak of the signal. By reason of this transformer action, therefore, the average voltage at the collector of the transistor remains at supply voltage when a signal is present.

It follows that in a Class A stage of this type, the current taken from the

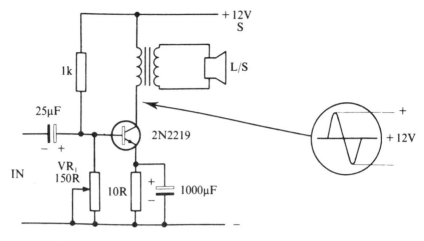

Figure 5.4   A single-transistor Class A amplifier

supply is constant whether a signal is amplified or not. Any variation in current flow from no-signal to full-signal conditions indicates that some non-linearity, and therefore distortion, must be present.

A circuit often used for Class B amplifiers is shown in Figure 5.5. It is known

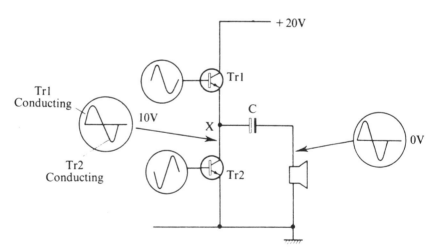

Figure 5.5   A single-ended push–pull or totem-pole circuit

as the *single-ended push–pull* or *totem-pole* circuit.

Two power transistors are connected in series, with their mid-connection (Point X in the figure) coupled through a capacitor to the load – which may be either a loudspeaker, the field coils of a TV receiver, or the armature of a servomotor. During the positive half of the signal cycle Tr1 conducts, so that the output signal drives current through the load from X to ground. During the negative half of the cycle, Tr1 is cut off and Tr2 conducts, so that the current now flows in the opposite direction, through the load and Tr2.

The coupling capacitor C_2 forms an essential part of the circuit. When there is no signal input, C_2 is charged to about half supply voltage i.e., the voltage at Point X. The voltage swing at this Point X, from full supply voltage in one direction to ground or zero voltage in the other, thus becomes at the load a voltage swing of identical amplitude centred round zero volts.

Summary

Transistor power amplifiers require transistors possessing low resistance and good ability to dissipate heat. In particular, the output transistor(s) must be tightly connected to a heat sink.

Class A stages, which pass large standing currents of steady value, need larger heat sinks than do Class B stages, which dissipate much less power in the transistors when no signal is applied.

Bias and feedback

Class A output stages need a bias system which will keep almost constant the standing current (i.e. the d.c. current with no signal input) flowing through them, despite the large temperature changes to which the standing current itself gives rise. One common method illustrated in Figure 5.6, is to use a silicon diode as part of the bias network. The diode should be a junction diode attacted to the same heat sink as the power transistor(s).

As a silicon junction is heated, the junction voltage (about 0.6V at low temperatures) which is needed for correct bias becomes less. A fixed-voltage bias supply composed of resistors would therefore over-bias the transistor as the temperature increased. A silicon diode compensates for the change in the base-emitter voltage of the transistor, since the forward voltage of the diode is also reduced as its temperature rises.

The Class A push–pull stage in Figure 5.6 uses two silicon transistors coupled to the load by a transformer. The input signal also is coupled to the stage by a transformer, which ensures that both transistors obtain the correct phase of signal. The use of a second transformer in such a circuit is often undesirable because it reduces the bandwidth and makes the amplifier less stable if negative feedback is used (because of the unpredictable changes of phase that occur in a transformer at extremes of frequency). A *phase-splitter* stage that uses transistors rather than a transformer can be used. Such a phase-splitter can use a transistor with equal loads in collector and emitter, or a long-tailed pair circuit (see Chapter 4) with unbalanced input and balanced output.

121

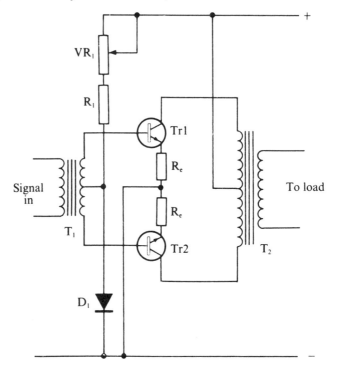

Figure 5.6   A silicon diode connected for use as bias control

The bias current for this circuit is taken to the centre-tap of the secondary winding of the phase-splitter (or *driver*) transformer T_1, and some additional stability is obtained by using the negative feedback resistors R_e in the emitter leads. Any increase in the bias current of the power transistors will cause the emitter voltage of both transistors to rise, so reducing the voltage between base and emitter and thereby giving back-bias to the input.

In both the circuits described in Figures 5.4 and 5.6 the bias is adjusted so that the current amount of steady bias current flows in the output stage. To adjust this bias current in the single-transistor stage requires that the collector circuit be broken to connect in a current meter, and that VR_1 be then adjusted so as to give the correct current reading. The push–pull circuit (Figure 5.6) can be set by measuring the voltage across the emitter resistors, R_e, and by adjusting VR_1 to give the correct value of voltage at these points.

A more elaborate circuit using the Class AB complementary stage is shown in Figure 5.7. In Class AB, each amplifier is biased to a value lying between those appropriate for either Class A or Class B. The output current in each stage thus flows for slightly more than half of each input cycle. The effect is to minimize cross-over distortion.

In this circuit, both a.c. and d.c. feedback loops are used. The d.c. feedback is

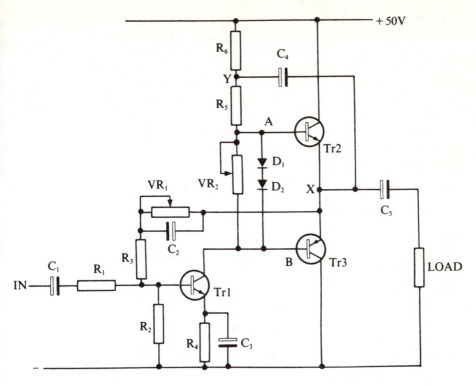

Figure 5.7   A complementary output stage of the totem-pole type, with driving stage and feedback connections

used to keep the steady bias current at its correct low value, and the a.c. feedback is used to correct the distortions caused by Class B operation – in particular the cross-over distortion.

Two bias-adjusting settings are needed in this circuit. The potentiometer VR_1 sets the value of the bias current in Tr1, so that the amount of current flowing through resistors R_2, R_3 and VR_2 is controlled. The potentiometer is adjusted so that the voltage at Point X is exactly half the supply voltage when there is no signal input. Potentiometer VR_2 controls the amount of current passing through the output transistors.

This type of output circuit is called a *complementary* stage, because it uses complementary transistors, one NPN and one PNP type, both connected as emitter followers. With the emitters of both Tr2 and Tr3 connected to the voltage at Point X, both output transistors are almost cut off when no signal is present. When VR_2 is correctly adjusted, the voltage drop between Points A and B is just enough to give the output transistors a small standing bias current (2 to 20mA) to ensure that they never cut off together. The value of this steady current is usually set low so as to keep cross-over distortion to a minimum –

with the actual value that recommended by the manufacturers.

A current meter must be used to check the value of bias current.

It will be seen that there are two a.c. feedback loops in this Figure 5.7 amplifier. Negative feedback, to improve linearity, is taken through C_2, and positive feedback is taken through C_4 to Point Y. This positive feedback, sometimes called *boot-strapping*, cannot cause oscillation because the feedback signal is of smaller amplitude than is the normal input signal at that point (an emitter follower has a voltage gain slightly lower than unity). It has, however, the desirable effect of decreasing the amount of signal amplitude needed at the input to Tr2.

Impedance matching

The ideal method of delivering power to a load would be to use transistors which had a very low resistance, so that most of the power (I^2R) was dissipated in the load. Most audio amplifiers today make use of such transistors to drive 8-ohm loudspeaker loads.

For some purposes, however, transistors offering higher resistance must be used or loads possessing very low resistance must be driven, and a transformer must therefore be used to match the differing impedances. In public address systems, for example, where loudspeakers are situated at considerable distances from the amplifier, it is normal to use high-voltage signals (100V) at low currents so as to avoid I^2R losses in the lines. In such cases the 8-ohm loud-speakers must be coupled to the lines through transformers.

The transformer ratio giving the best transfer of power is expressed by the formula:

$$N = \sqrt{\frac{\text{Output impedance of amplifier}}{\text{Impedance of load}}}$$

where N is the ratio of the number of turns in the primary winding of the transformer to the number of turns in the secondary winding.

Example. A power amplifier stage operates with a 64-ohm output impedance. What transformer ratio is needed for maximum power transfer to an 8-ohm load?
Solution. $N = \sqrt{64/8} = \sqrt{8} = 2.8$. A 3:1 step-down transformer would therefore be used.

Exercise 5.1

Set up the circuit shown in Figure 5.8, using a multi-ratio output transformer. Tr1 can be either a 2N3053 or a BFY50, Tr2 a 2N3055 and D_1 a 1N4001.

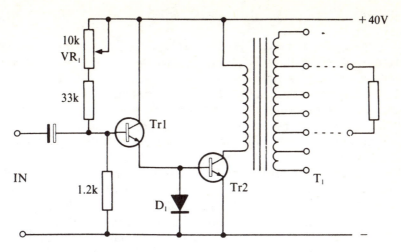

Figure 5.8   Circuit for Exercise 5.1

With a 40V supply, adjust VR$_1$ so that the standing d.c. current flow through Tr2 is 50mA. Connect the 8-ohm resistor between two of the taps of the secondary of the transformer and connect the signal generator to the input. Connect an oscilloscope across the 8-ohm resistor.

Applying a 400 Hz signal, adjust input signal voltage so that a 2V peak-to-peak signal is observed across the 8-ohm resistor. Now, without altering the signal generator settings, switch off the amplifier, change the transformer connections so as to alter the ratio, and switch on again. Measure the output voltage.

Repeat the procedure so that the output voltage is measured for every possible pair of secondary tapping points. What transformer ratio gives the maximum signal output? Is the setting critical?

Exercise 5.2

Connect into the power amplifier of Exercise 5.1 (or any other suitable amplifier), an 8-ohm load resistor, and adjust the signal generator to give a 2V peak-to-peak output across it. Vary the frequency of the signal generator, first down to the frequency at which the voltage output is 1.4V (0.707 × 2V), or '3 dB down', and then upwards to the frequency at which the output is again 1.4V, or '3 dB up'. Note these 3 dB frequencies.

It will be seen that at each of these '3 dB' frequencies the power output of the amplifier is half as much as it is at 400Hz (0.707 × 0.707 ≃ 0.5). The frequencies are therefore known as the *half-power points*. Note the power bandwidth of the amplifier, which is the frequency range between the half-power points.

Summary

Bias systems for power output stages must keep the bias current steady over a wide range of temperatures. This is more difficult in Class A stages, because they operate with large bias currents.

Signal coupling from amplifier to load is generally by capacitor or transformer, but sometimes direct.

When a transformer is used to couple stages of unequal resistance, the transformer ratio must be chosen so as to match correctly the impedances of the two stages.

Both d.c. and a.c. feedback can be used in power amplifiers. D.c. feedback will be used to stabilize the bias, particularly against variations caused by changes of temperature. A.c. feedback is used to stabilize the voltage gain, often to a very low value, so as to keep distortion to a minimum. Remember that a power amplifier seldom needs to possess any voltage gain; its main purpose is to provide current gain while maintaining the voltage level of a signal.

The power output that can be obtained from a given design depends to a very large extent on the ability of the heatsink to dissipate the heat from the output transistors (or IC).

Faults

Faults in output stages are usually caused by over-dissipation of power, which in turn can be caused by over-loading or over-heating. An output stage can be over-loaded, when capacitor coupling is used, by connecting a load having too low a resistance. The usual result is to burn out the output transistor(s).

In most Class AB totem-pole circuits, even the most momentary short circuit at the output (caused for example, by faulty connections) will cause the output transistor (usually Tr2 in Figure 5.7) to burn out if a signal is being amplified.

Excessive bias currents can often be traced to the failure of a diode in the bias chain, or to a burn-out of the biasing potentiometer in a totem-pole circuit.

Unexpected clipping of the output can be caused by failure of the bootstrap capacitor (C_4 in Figure 5.7), or by a fault in the bias resistors which has caused the voltage at Point X to drift up or down.

IC power amplifiers are widely used, replacing power stages which use separate (discrete) transitors. Unlike op-amps, however, power amplifier ICs take no standardized form, though many use the same scheme of connections. Figure 5.9 shows a circuit diagram (courtesy of RS Components Ltd.) for the TDA2030 power amplifier used with a single 15V power supply. The IC is manufactured on a steel tab which can be bolted to a heat-sink for efficient cooling.

The circuit is very similar to that of a non-inverting op-amp voltage amplifier,

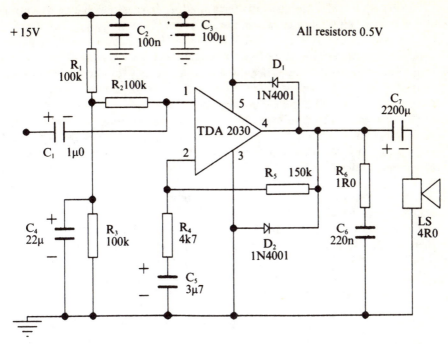

Figure 5.9   A power amplifier circuit, courtesy of RS Components Ltd, using the TDA 2030 IC

but the practical layout of the circuit has to be designed so that the decoupling capacitors (C_2 and C_3) are very close to the IC. The electrolytic capacitor C_3 needs to be bypassed by a plastic dielectric capacitor because the impedance of an electrolytic capacitor rises at the higher frequencies. At these frequencies, C_2 performs the decoupling in place of C_3. The series circuit of R_6 and C_6 is designed to maintain stability and is called a Zobel network – it is also widely used in discrete transistor amplifiers. This circuit will deliver up to 13W of audio output into the 4Ω loudspeaker.

Power supplies – regulation

The simplest possible power-supply circuits consist of rectifiers and smoothing capacitors, which have already been covered in Chapter 2. In these circuits, the reservoir capacitor supplies the current to the load during the time when the diodes are cut off, and any ripple on the supply is reduced by filtering. Such circuits are adequate for many purposes, but they are too poorly regulated for use in circuits intended for measurement, computing, broadcasting or process control.

The *regulation* of a circuit is the term used to express the change of output voltage caused either by a change in the a.c. supply voltage or by a change in the output load current. A well regulated supply will have an output voltage whose value is almost constant; the output voltage of a poorly regulated supply will change considerably when either the a.c. input voltage or the output current changes. The table on page 35 shows how the voltage of a smoothed supply changes from no-load to maximum-load condition.

The change in output voltage caused by changes in the a.c. supply voltage is usually less great. A 10% change in the a.c. supply will also change the output voltage of a simple power supply by about 10% – the two percentage changes being almost identical.

Example. What will be the change in a 10V supply when the a.c. supply voltage changes from 240V to 220V?

Solution. The a.c. voltage change is 20V down on 240. The percentage change is therefore $20/240 \times 100 = 8.3\%$. Since 8.3% of 10V is 0.83V, the 10V supply will drop to $10 - 0.83V = 9.17V$.

The effect of changes in the load current (which are themselves caused by changes in the load resistance) is more complicated, because there are two such effects (see Figure 5.10). The first is brought about by the resistances of the rectifiers, the transformer windings and any inductors which may be used in

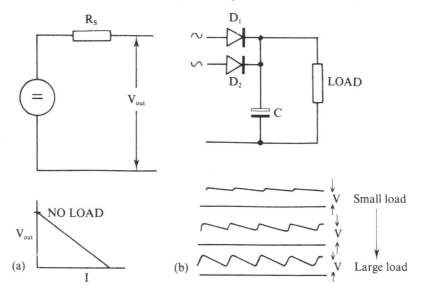

Figure 5.10   The causes of poor regulation
(a) The combined resistances of circuit rectifiers, transformer windings, etc. and (b) the charge and discharge of the reservoir capacitor

filter circuits, and by the internal resistance of the power supply. By Ohm's law, a change of current flow through such resistances must cause a change of voltage.

The second effect lies in the voltage drop which takes place across the reservoir capacitor as more current is drawn from it. This voltage drop (V) can be quantified by the equation:

$$V = \frac{I \times t}{C}$$

where I is the load current in amps, t the time in seconds elapsing between one charge from the reservoir and the next, and C the capacitance in farads of the reservoir.

Example. By how much will the voltage across a 220μF capacitor drop when 0.2A is drawn from a full-wave rectifying circuit?
Solution. In a full-wave rectifying circuit, the time between peaks is 10ms.

Substituting the data in the equation $V = \frac{I \times t}{C}$,

$$V = \frac{0.2 \times 0.01}{220 \times 10^{-6}} = 9V$$

A 50V supply, for example, would give an output which drops to 41V between peaks, and a 9V peak-to-peak ripple at 100Hz would be present.

Summary

The simple rectifier-reservoir power supply gives poor regulation. Its output voltage is proportional to the a.c. input voltage; and the effect of the load current on the circuit resistances and on the charge on the reservoir capacitor causes sharp voltage drops and a large amount of ripple at maximum d.c. current flow.

Regulator circuits

A regulator circuit connected to a rectifier/reservoir unit ensures that the output voltage is steady for all designed values of load current or of a.c. supply voltage. Such a circuit can only do its job, however, if the rectifier/reservoir unit itself is capable of supplying the required output voltage (measured from the minimum of the ripple wave) under the worst possible conditions, i.e., when a.c. supply voltage is minimum and load current is maximum. The regulator will then prevent the output voltage from rising above this set value even when load current is small or the a.c. supply voltage high.

Two types of regulator are used – a series circuit and a shunt circuit. The simplest example of the shunt type is a Zener diode regulator shown in Figure 5.11. In this type of circuit, the amount of current drawn from the power supply

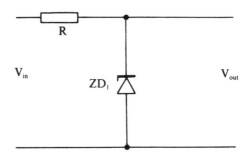

Figure 5.11   A simple Zener-diode regulator

is constant. When load current is maximum, the regulator circuit current is minimum. When the load takes its minimum current, the regulator circuit takes its maximum current. The arrangement is therefore such that load current plus regulator current is always a constant value. The constant current flowing through R produces a constant voltage drop across it. So if the input voltage remains unchanged, the output voltage will remain constant also.

If the supply voltage changes, the total current will change to a new value, but the action of the Zener diode ensures that there is no change in output voltage.

Some care has to be taken, however, over the power ratings of the Zener diode and of its supply resistor. The maximum dissipation of the Zener diode in the circuit is: Zener voltage × maximum current – with dissipation in milliwatts if the current is measured in mA.

The maximum dissipation of the resistor is given by:

(Unregulated voltage – Zener voltage) × maximum current.

Example. A 5.6V Zener diode is used to supply a load which takes a maximum current of 15mA. If the minimum desirable Zener current is 2mA, and the unregulated voltage is 12V, find (a) the value of series resistance which must be used, (b) the maximum Zener dissipation and (c) the maximum dissipation in the resistor.

Solution. With 15mA flowing through the load and 2mA through the Zener diode, total current is 17mA. The voltage across the resistor is $(12 - 5.6 =)$ 6.4V, so that (a) the required resistance value, using Ohm's law, is 6.4/17 = 0.376k or 376 ohms. In practice a 330-ohm resistor would be used, making the total current 6.4/.330mA = 19.4mA. (b) At a current of 19.4mA, the dissipation in the Zener diode is $5.6 \times 19.4 = 108.6$mW; and (c) the dissipation in the resistor is 6.4 × 19.4 = 124.2mW.

When the circuit requires that much larger currents than about 20mA be passed through the stabilizer, the circuit shown in Figure 5.12 can be used. In

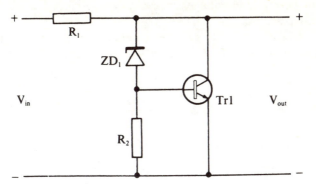

Figure 5.12   The 'Amplified-Zener' regulator circuit

this circuit, the Zener diode supplies the base current for a power transistor connected as a shunt regulator.

For the transistor to conduct, its base voltage must be about 0.6V higher than its emitter voltage, so that the voltage at its collector must be about Zener voltage +0.6V. Any rise in the collector voltage would cause an equal rise of voltage at the base, because the Zener diode keeps the voltage between base and collector constant. A larger base voltage would cause a much higher collector current (remember that a rise of 60mV in base voltage causes collector current flow to increase by ten times), and the voltage drop across the series resistor would thus restore the correct operating conditions.

This negative feedback arrangement is sometimes called an *amplified Zener circuit*.

Both the straightforward Zener diode and the amplified Zener circuits have the additional advantage that they greatly reduce the amount of hum ripple from the power supply, provided that the amount of current being taken from the supply does not cause the minimum voltage to drop below the regulated output voltage.

The output voltage will not change noticeably when the a.c. supply voltage changes, provided that the voltage does not fall below the level needed to keep current flowing through the Zener diode.

Exercise 5.3

Using a simple rectifier-plus-reservoir power supply, plot regulation curves. Supply the a.c. input through a Variac auto-transformer, and read off the input a.c. voltage with an a.c. voltmeter – making sure that all connections are insulated before switching on.

Now use a d.c. voltmeter to measure the output voltage of the power supply. With no load, measure output voltage for a.c. input voltages ranging from 90% to 110% of normal line voltage, entering your results on a graph of output voltage plotted against input (a.c.) voltage.

Now set the input a.c. voltage to normal line voltage, and add the load and ammeter circuit of Figure 5.13. Note the value of output voltage for currents of

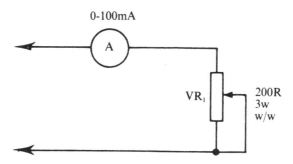

Figure 5.13   Circuit for Exercise 5.3

0.10mA, 50mA, 75mA and 100mA, and plot a graph of output voltage against load current.

These two graphs form the regulation graphs for the power supply.

Exercise 5.4

Make up the regulator circuit shown in Figure 5.12 with the following values of component: $R_1 - 33R$, 1W (W/W – or wire-wound); $ZD_1 - 5.6V$; $R_2 - 330R$; Tr1 – 2N3055; $V_{in} - 10V$. Connect this circuit to the power supply used in Exercise 5.3, and draw another set of regulation curves for the regulated output. Measure also the ripple at 100mA load current (a) across the reservoir capacitor and (b) across the load.

The other way of regulating the output of a power supply is by means of a series regulator in which a transistor is connected between the supply and the load, as indicated in Figure 5.14.

When this type of regulator circuit is used, the transistor takes only as much current as the load – unlike the shunt regulator which takes its maximum current just as the load takes minimum current. The base voltage of Tr1 is held constant by the regulator action of ZD_1 and R_1. The emitter voltage, and so the conduction, of Tr1 depends on the load voltage. If the demand for load current increases, the output voltage will tend to fall, increasing the forward bias of Tr1. This allows it to pass more current to meet the demand.

If the requirement for load current falls, this effect will be reversed.

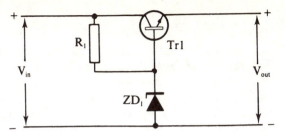

Figure 5.14   The series regulator

Exercise 5.5

Make up the simple series regulator circuit of Figure 5.14 with the following component values: R_1 – 330R; ZD_1 – 5.6V; Tr1 – 2N3055; V_{in} – 10V. Connect this circuit to the power supply previously used, and draws a set of regulation curves for the stabilised circuit.

Measure also the ripple voltage at maximum load current (a) across the reservoir capacitor and (b) across the load.

More elaborate series regulator circuits use comparator amplifiers to drive the series transistor.

In the circuit pictured in Figure 5.15, the positive (in-phase) input to a

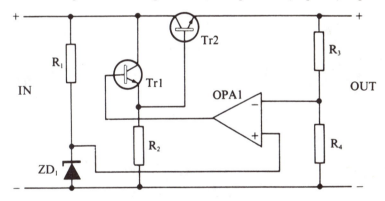

Figure 5.15   A series regulator circuit containing a comparator amplifier

comparator amplifier labelled OPA1 is connected to a Zener diode so that its voltage is fixed. The negative (anti-phase) input is connected to the output of the power supply through a potential divider. The output of the comparator drives the series transistor so that any drop in the output voltage is at once counteracted by the flow of more current through the transistor.

Excellent regulation can be provided by such a circuit, which is nowadays available complete in integrated form.

Fault-finding

Failure in regulator circuits is usually caused by excessive dissipation in the main transistor, be it shunt or series as the case may be.

In the series circuit the over-dissipation will have been caused by an excessive load current, unless the regulator is protected against short circuits.

In a shunt circuit, excessive load currents will not directly damage the main transistor; but it will often impair the series resistor so that, when the load is removed, excessive current flow will then pass through the shunt transistor.

An o/c Zener diode will cause a shunt regulator to cease conducting. Its effect on the series circuit is the opposite, in that the output voltage will rise to the level of the unregulated supply.

An s/c Zener diode will cause the shunt circuit to pass excessive current. It will cause the series circuit to cut off.

Note carefully that all power transistors used in regulator circuits must be bolted to heat sinks of adequate size.

Exercise 5.6

Use the rectifier-stabilizer circuit of Figure 5.16 to assess the effect of faults,

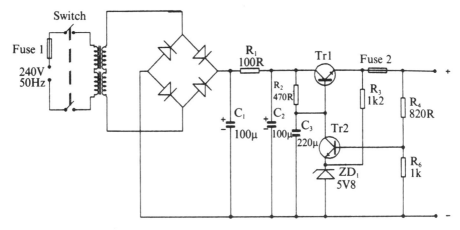

Figure 5.16   A rectifier-stabilizer circuit using, unusually, discrete transistors so that the effect of faults can be demonstrated

using either faulty components or simulating faults. For the circuit as shown, measure and tabulate the following:

(a) current flowing with no additional load
(b) d.c. output voltage

(c) d.c. output voltage when a 1k load is connected
(d) current flowing when the 1k load is added.

Now note the voltage levels at each terminal of each transistor, 6 voltage readings in all. Repeat these measurements when the following faults (one fault at a time) are introduced.

(a) Tr1 b–e o/c
(b) ZD_1 o/c
(c) R_3 equal to 22k
(d) Tr2 b–e o/c.

Explain the effect each fault has on the voltage levels.

Voltage-doubler circuits

There are a few circuits which call for a high-voltage, low-current supply in which poor regulation is acceptable. Rather than wind a transformer especially to provide the high voltage, a *voltage multiplier* circuit (of which the *voltage doubler* is the simplest example) is often used instead.

On the negative-going half of the voltage cycle shown in Figure 5.17, C_1 is

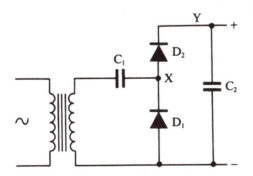

Figure 5.17   A voltage-doubler circuit

charged by current through D_1, so that Point X is at a d.c. voltage equal to the peak voltage of the a.c. wave.

At the peak of the positive-going half-cycle, the peak inverse voltage across D_1 is equal to twice the peak voltage (the previous peak charge, plus the peak a.c. value). This causes C_2 to charge to the same level through D_2.

At line (supply) frequencies, the capacitor C_1 must be of large value, and must, of course, be rated at the full d.c. voltage. At higher frequencies, smaller values of capacitance can be used.

135

Multiplier circuits of this type are commonly used to supply the high voltages required by the colour tubes in colour TV receivers.

The circuit shown in Figure 5.18(a) is effectively two voltage-doublers

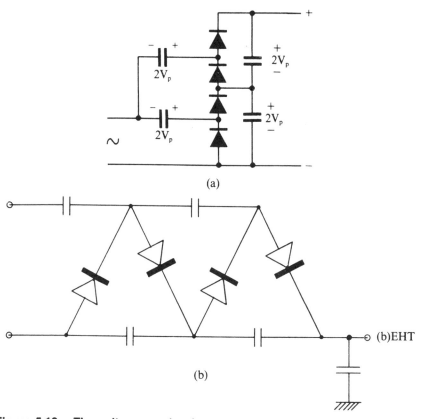

(a)

(b)

Figure 5.18   The voltage-quadrupler
(a) circuit (b) alternative configurations

connected in series, while Figure 5.18(b) shows the so-called *Cockcroft-Walton multiplier*. Circuits such as these are commonly used to obtain voltages as high as 25kV for the final anode electrode of the tube.

Summary

Regulator circuits are used to keep steady the output voltage from a power supply, at a value equal to the lowest voltage output from the reservoir capacitor on full load and with minimum a.c. input voltage.

Either shunt or series regulators (sometimes also called *stabilizers*) can be used. Either type of circuit provides great improvement in regulation and control of ripple; but the series circuit is in more common use, apart from the simple Zener diode regulator.

In high-voltage supplies, the use of voltage-multiplier circuits provides an alternative to relatively expensive high-voltage transformer windings.

Components

The components used in power supplies must be adequately rated for the voltage and current levels that will be used. Electrolytic capacitors can be used for low-voltage smoothing applications, but they should not be used for high-voltage supplies, particularly above 500V. For such applications, plastic dielectric capacitor must be used and for some specialized applications, oil-filled paper capacitors. Resistors are subject to a maximum voltage rating, often 350V or less, and where higher voltage uses are essential, chains of resistors connected in series must be used to keep the voltage drop across each individual resistor within the limit.

Power supplies which operate at low voltages usually deliver large currents, and computer power supplies in particular will supply 20A or more at +5V. Electrolytic capacitors used for smoothing such supplies (which are nearly always switch-mode supplies, see later) must be rated to have low equivalent series resistance and be able to withstand high ripple currents. Some supplies make use of a series resistor for monitoring current, and this resistor will have a very low resistance value and cannot be replaced with a conventional resistor.

Controlled power supplies

Power supplies designed to supply d.c. or a.c. at high power or high voltage need to be of rather different design. Many of them use thyristors as the controlling semiconductors.

A thyristor, it will be recalled, can be triggered on but not off. In the design of a power supply stabilised by thyristors, current shut-off is achieved by feeding the thyristor or thyristors either with a.c. or with uni-directional pulses having a zero or a negative voltage peak. If a d.c. output is required, a diode must be placed in circuit between the thyristor and the smoothing capacitor (Figure 5.19) to enable the thyristor to be switched off.

Two different types of control can be achieved with the aid of thyristors. First, in a *phase control system*, the level of average current flow is controlled by switching on the thyristor at different parts of the cycle (Figure 5.20). In such a system, the thyristor conducts on each positive cycle, but the phase of the gate signal varies according to the output required. A disadvantage is the large current surge which arises when the thyristor is switched on, causing radio-

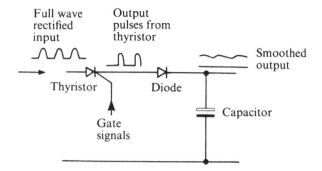

Figure 5.19   A diode placed between thyristor and capacitor

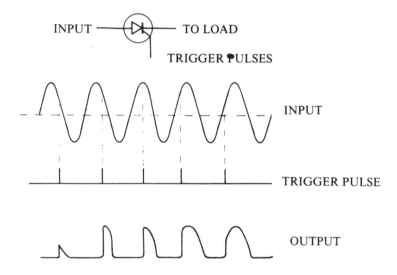

Figure 5.20   A thyristor in a phase control system

frequency interference (RFI). Phase control systems therefore require chokes and suppression capacitors to reduce the radiation of interference and also to prevent interference pulses from other thyristor-controlled equipment from affecting the thyristor gate.

The second type of control system, known as *zero-voltage switching* or *burst-firing system*, can be used only when the load is a heater or a motor having a large flywheel. In a zero-voltage switching system (see waveforms in Figure 5.21), the thyristor is only switched on when the voltage between its anode and cathode is zero, so only little interference can be generated. A few complete cycles of the waveform are allowed to pass, then the gating pulses are removed. The waveform to the load thus consists of a few complete cycles, followed by

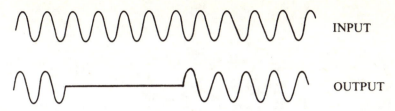

INPUT

OUTPUT

Figure 5.21   A thyristor in a zero-voltage switching system

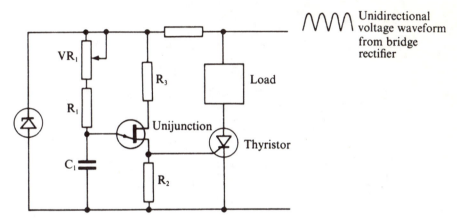

Unidirectional voltage waveform from bridge rectifier

VR₁

R₃

Load

R₁

Unijunction

Thyristor

C₁

R₂

Figure 5.22   A typical thyristor power controller

zero input. This method of control can therefore be used with loads which have a long time-constant.

Figure 5.22 shows a circuit in which a thyristor is used in the phase-control mode of operation to control average current flow through a load. On each half-cycle, a thyristor remains non-conducting until C_1 has charged sufficiently to fire the unijunction. The current pulse through R_2 then fires the thyristor, and current passes through the load for the remainder of the half-cycle.

If the load is connected in the line leading to the bridge rectifier, an a.c. load can also be controlled by this technique.

Figure 5.23 shows the circuit diagram of a stabilized power supply using thyristors – a circuit which has been used to power a Thorn Electrical TV receiver.

C712 is a capacitor which is charged from the (separately stabilized) +25V line through R722 and R723. The charging current is controlled by VT706, and its value is set by the voltage reaching the base of this PNP transistor from the 180V line through R724, R725 (the voltage preset) and R726.

The 180V line is the supply which is stabilized by the action of the thyristor. If this voltage is low, making the base of VT706 more negative relative to its

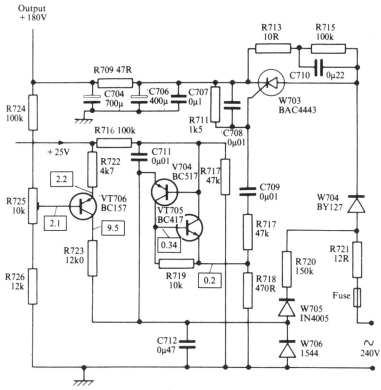

Figure 5.23   A thyristor-controlled power supply

emitter voltage, the collector current will increase and C712 will charge more rapidly. If the 180V line is at too high a voltage, however, VT706 will be biased back and C712 will charge more slowly.

C712 is discharged on each cycle by the mains voltage applied through R720 and the diode W705, with W706 acting as a limiter.

The two transistors VT704 and VT705 are so connected as to act like a unijunction. As the voltage across C712 builds up, therefore, the voltage at the emitter of VT704 will eventually cause both transistors to pass current, so discharging C712 through R718 and causing a pulse of voltage across R718 which fires the thyristor 2703.

The higher the voltage of the output line, the later in the cycle the thyristor fires and the more the output drops, so restoring the correct voltage. In the circuit shown, the figures in rectangles indicate the steady bias voltages.

Figure 5.24 shows part of a circuit which is used for controlling the speed of a shunt-wound d.c. motor. The diode D_1 is known as a *flywheel* (or *freewheel*) diode. Its function is to permit current to continue flowing freely in the circuit when the thyristor is not conducting.

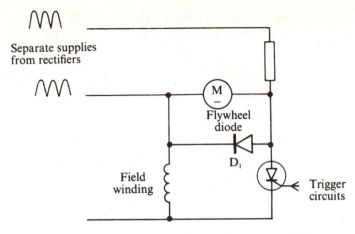

Figure 5.24   Control circuit (part only) for a shunt-wound d.c. motor

Switched mode power supplies

The most common configuration of linear-regulated power supply consists of a mains frequency transformer and rectifier, together with a transistor series regulator. This latter simply behaves as a controlled series resistance to stabilize the output voltage. Such systems suffer from several serious disadvantages:

1   They are most inefficient. It is unusual to find that more than 35% of the input energy reaches the load. The remainder is dissipated as heat.
2   The mains transformer is invariably large. Its size tends to be inversely proportional to the operating frequency.
3   The reservoir and smoothing capacitors need to be large to keep the ripple amplitude within acceptable bounds.
4   Because the series transistor (or transistors) is operated in the linear mode they must be mounted on large heat sinks.

If the operating frequency can be increased significantly, both the transformer and the filter capacitors can be reduced in size. If the series transistor can be operated either cut-off or saturated, its dissipation will be reduced. The power supply can then be made more efficient. Such operation can be achieved using a switched mode power supply (SMPS). These circuits can operate with efficiencies as high as 85%.

The basic principle of the SMPS is shown in Figure 5.25. When the switch is closed, current flows through the inductor or choke L to power the load and charge the capacitor C. When the switch is opened, the magnetic field that has been built up around L now collapses and induces an emf into itself to keep the current flowing, but now through the flywheel or freewheel diode, D. The

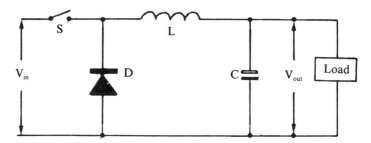

Figure 5.25   Basic principle of switched mode power supply

voltage across C now starts to fall as the load continues to draw current. If the switch is closed again the capacitor recharges.

The duty cycle or switching sequence is shown in Figure 5.26 together with

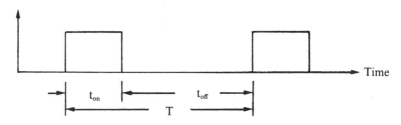

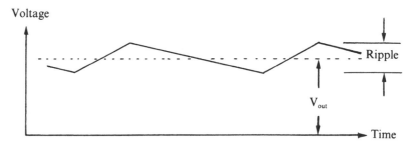

Figure 5.26   Output voltage and switching action

the output voltage V_{out} that it produces. Increasing the 'on-period' will increase V_{out} whose average level is given by $V_{in} \times \frac{t_{on}}{T}$. V_{out} can be regulated by varying the mark to space ratio of the switching period. Any wanted ripple can be filtered off in the usual way.

The typical SMPS whose block diagram is shown in Figure 5.27, consists of a mains rectifier with simple smoothing whose d.c. output is 'chopped' or switched at a high frequency, the switch being a transistor, which for TV applications is commonly driven at 15.625kHz. For industrial applications, the

142

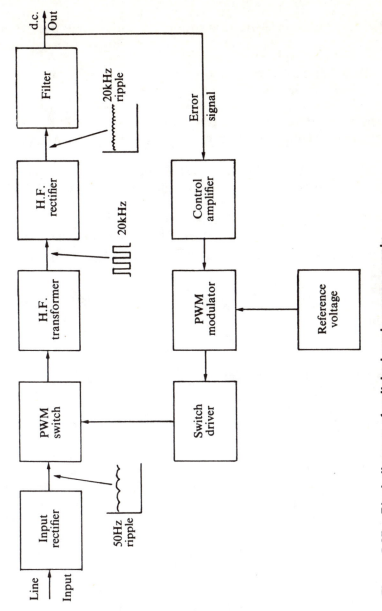

Figure 5.27   Block diagram of switched mode power supply

143

switching frequency is usually in the order of 20 to 25kHz. The chopped waveform is applied to the primary circuit of a high-frequency transformer that uses a ferrite core for high efficiency. The signal voltage developed at the secondary is rectified and filtered to give the required d.c. output. This output is sensed by a control section that makes a comparison with a reference voltage to produce a correction signal that is used to modulate the *mark-to-space* ratio of the switching circuit to compensate for any variation of output voltage. This action is effectively pulse-width-modulation.

The ripple frequency of 50Hz at the input has been translated to a frequency of 20kHz at the output. Thus the frequency has been increased by a factor of 400, so that the smoothing and filter capacitors can be reduced in value by the same ratio.

The circuit can be operated from a battery or any other d.c. input, so that the SMPS becomes a device for converting d.c. from a high voltage to one at a lower level.

The SMPS does tend to generate more radiated and line conducted noise than a linear supply. This can be reduced to acceptable levels by using:

1 Mains input filters balanced to earth to give rejection of the switching frequency.
2 Suitable design of output filter.
3 Electro-static screen between primary and secondary of the mains transformer.
4 Efficient screening of the complete unit.

Excess voltage trip (crow-bar)

Figure 5.28 shows the layout of an excess voltage protection circuit. Tr1 is the switcher transistor, D_2 is typically a 72-volt Zener diode and Th_1 a silicon-controlled rectifier. When the output voltage is held at its normal level of, say, 65 volts, D_2 is reverse-biased and non-conducting. There is, therefore, no voltage developed across R_1, and the gate voltage of Th_1 is zero.

If the output voltage rises above 72 volts, D_2 conducts, draws current through R_1 and the gate voltage of Th_1 rises to trigger it into conduction. Thus a short circuit is applied across the supply input, which can either blow the mains input fuse or trigger a thermal cut-out to break the circuit. The effect of causing the conduction of Th_1 is therefore like placing a very low resistance (the crow-bar) across the supply rails under over-voltage conditions.

Excess current protection (dynamic current trip)

The circuit shown in Figure 5.29 provides the SMPS unit with protection against excess current demands. L_1 offers a high impedance to the switching frequency current and is therefore by-passed by C_1 and R_1. There is thus an a.c. voltage developed across R_1, which could cause Th_1 to conduct if the negative excursion exceeds the gate voltage as set by R_5. If under fault conditions, the current drawn from the 65-volt line rises, there will be a corresponding increase in the a.c. voltage across R_1. On the negative excursion, Th_1 will therefore be fired to short out the switching pulses and cause the 65-volt line voltage to fall so that Th_1 again becomes a high resistance and the circuit resets. If the fault condition still persists, the process of 'trip and reset' repeats at the switching frequency allowing only a limited current to be drawn.

Exercise 5.7

The behaviour of SMPS can be effectively studied using a suitable TV receiver. Although the power provided is relatively small compared with many industrial units, the TV receivers are fairly readily available on the surplus market. One range that has been found to be particularly suitable is the Thorn 3000/3500 series. These provide an opportunity to study power supply behaviour as well as over voltage current protection. Over current conditions can be simulated using a laboratory-type resistor (rheostat) and a mains voltage lamp (100W). Over voltage can be provided by using a variac auto-transformer. Alternatively, switched-mode power supply kits are available from LAMBDA Electronics Co. which provides 5 volts at 5 amps output. These not only provide valuable construction exercises and are useful as TTL type power supplies, but they can also be used to study the regulation and ripple characteristics of SMPS.

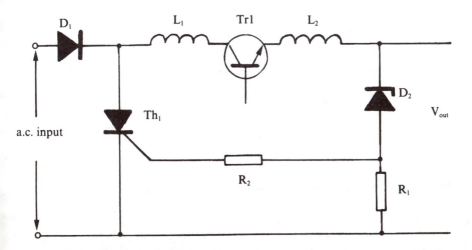

Figure 5.28   Excess voltage trip circuit

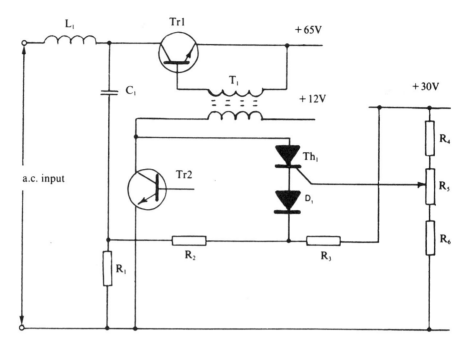

Figure 5.29   Excess current protection circuit

Inverter and converter power supplies

Although inverters are designed to convert d.c. power into a.c. power, they have some features in common with converters that are used to transform d.c. energy from a low-voltage level into d.c. at a higher voltage. The block diagram of Figure 5.30 shows some of these common features. An oscillator, usually

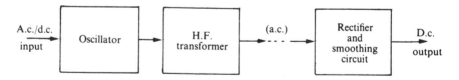

Figure 5.30   Block diagram of inverter/converter circuits

running at an ultra-sonic frequency (typically around 20kHz) is powered from a d.c. source. This oscillator drives a transformer to provide either an a.c. output in the case of an inverter, or to power a rectifier/smoothing circuit for converter applications. It is important to note that in each case the total output power

146

must be lower than that supplied from the d.c. source due to the energy losses in the transformer, etc.

Line filters and interference reduction

When using mains-powered equipment it is important that the minimum of noise and interference generated should be fed back into the mains. This is particularly important where data-processing equipment is operating in close proximity to high-current industrial plant. Power supplies of the SMPS or oscillator types in particular should have their inputs well filtered to prevent pollution of the mains supply. Typical filter circuits and component values are shown in Figure 5.31 (a) and (b). Inductors and capacitors used in these

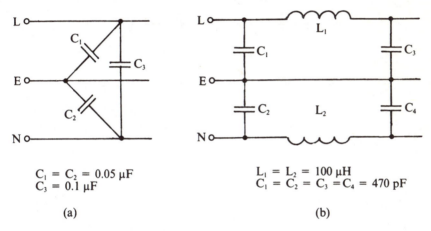

$C_1 = C_2 = 0.05\ \mu F$
$C_3 = 0.1\ \mu F$

$L_1 = L_2 = 100\ \mu H$
$C_1 = C_2 = C_3 = C_4 = 470\ pF$

(a) (b)

Figure 5.31 Line interference filters

applications should have a working voltage and/or current ratings under fault conditions. The characteristics of such filters are low pass, with zero attenuation at 50Hz and at least 30 dBs over the frequency range 150kHz to 50MHz. Data-processing equipment is particularly susceptible to mains-borne interference. Data may be corrupted by the high voltage transients induced from inductive loads on the mains supplies. Fortunately this problem is fairly easily solved by wiring devices called *varistors* across each pair of the mains supply wiring.

Varistors are usually made of zinc oxide or silicon carbide, with the former being often preferred because of its faster response. The characteristic of a varistor is shown in Figure 5.32. The device has a high resistance below some critical voltage, but above this, it rapidly conducts to short circuit any large over-voltage condition. The working voltages of these devices, which can be used for both a.c. and d.c., range from about 60 volts up to about 650 volts.

Mains power outlets and mains lead plugs are now available for low power

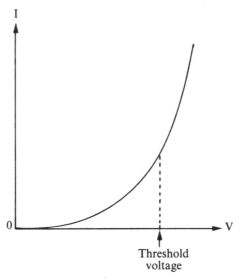

Figure 5.32   General characteristic of a varistor

(less than about 2kW) applications. These contain both filters and surge suppressors. They are particularly suited to the mains supplies for mini- and micro-computer installations.

Summary

Switch mode power supplies have the following advantages:

- small bulk
- low dissipation
- high-current outputs
- no heavy transformers
- no elaborate filters
- stabilized output

and the following disadvantages:

- more elaborate circuit
- more RF generated.

Rechargeable cells

Cells of the nickel-cadmium type are much used in portable electronic equipment because they can be easily recharged. The recharging, however, must be done at a constant rate of current flow, the value of which must be that specified by the manufacturer of the cells. The cells being sealed, excessive current flow can cause a build-up of gas which will eventually cause the cells to rupture explosively.

A basic constant-current circuit is illustrated in Figure 5.33 An LED is used

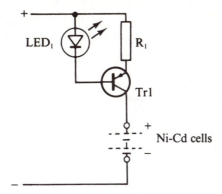

Figure 5.33   A basic nickel-cadmium battery charger

to maintain a constant voltage between the base and the emitter of transistor Tr1 – with the result that a constant current, whose size is controlled by the value of R_1, will flow in the collector circuit when the cells are connected into place. The charging current remains constant irrespective of how many cells are connected in series, provided that the supply voltage is high enough to permit current to flow at all.

A more elaborate circuit of the same kind is shown in Figure 5.34. Here the

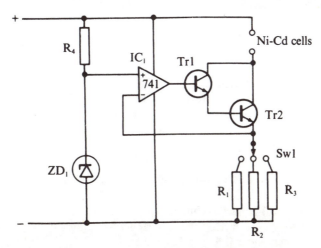

Figure 5.34   A variable-rate constant-current battery charger

149

current flowing through whichever of the resistors is selected by Sw1 causes a voltage drop which is applied to the ($-$) input of the Type 741 op-amp. Since the op-amp's ($+$) input is set to 5.6V by the Zener diode ZD_1, the output of the 741 will so bias Tr1 that the ($-$) input of the 741 will also be at 5.6V. By Ohm's law, the amount of current flowing through the selected resistor will be 5.6/R, where R is the value of the resistor selected by Sw1.

Constant-current charging systems can be further elaborated by adding circuits to detect over-voltage. Charging will then automatically stop when the cells are fully charged, and an indicator light will be switched on.

Multiple-choice test questions

1 The main action of a power amplifier stage is to provide:
   (a) current gain
   (b) reduced distortion
   (c) voltage gain
   (d) reduced bandwidth.

2 A transformerless push–pull power amplifier must use:
   (a) a stabilized supply
   (b) a phase-splitter stage
   (c) high-resistance transistors
   (d) complementary transistors

3 The reservoir capacitor of a power supply does not:
   (a) stabilize the output
   (b) reduce r.f. interference
   (c) reduce ripple
   (d) increase the d.c. level.

4 A series regulator transistor:
   (a) has the same voltage drop across it as the load
   (b) carries the same current as the load but with a varying voltage drop
   (c) dissipates the same amount of power as the load
   (d) has a constant voltage drop across it.

5 A typical switched mode power supply uses 40kHz switching. This ensures that:
   (a) the output will be well stabilized
   (b) smoothing will require only small capacitor values
   (c) output resistance will be low
   (d) no transformer will be required.

6   A TV power supply uses thyristors in a bridge rectifier circuit for mains
    voltage and mains frequency. This type of circuit:
    (a) allows much smaller smoothing capacitors to be used
    (b) always requires a transformer to be used
    (c) allows the voltage to be stabilized by controlling the thyristors
    (d) permits the chassis of the receiver to be isolated.

6 LCR circuits

Summary

Transient currents in inductors and capacitors. Back–e.m.f. Charging current. Time constant. Integration and differentiation. Reactance. Phase angle. Phasor diagrams. Impedance. Filters. Resonance. Damping.

D.c. circuits

In any circuit receiving a steady d.c. supply voltage, a resistor will act to control current flow. The equation to determine the value of that flow is Ohm's law in its $I = V/R$ form. If there is neither inductance nor capacitance in the circuit, the value of current given by the Ohm's law equation will start to flow at the instant the voltage V is switched on, and will continue to flow for as long as that voltage continues to be applied. If the value of the voltage is changed the value of the current will change also – but the strict relationship $I = V/R$ will always be preserved.

The circuits shown in Figure 6.1, however, contain either capacitors or inductors as well as resistors. The addition of such components to a d.c. circuit causes the value of current flow in the circuit to vary for a short time after the circuit voltage is switched on, before finally settling down at the value which the use of Ohm's law would predict. It is important to distinguish between the *steady* current, which is the final value of current, and the *transient* current, which is the changing value flowing just after the circuit has been switched either on or off.

152

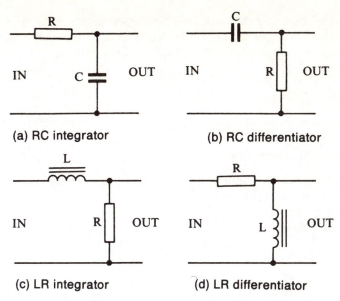

(a) RC integrator   (b) RC differentiator

(c) LR integrator   (d) LR differentiator

Figure 6.1   Some circuits in which transient currents can flow

The time during which the transient current flows is important. It is measured by a quantity called the *time constant* of the circuit in question, and it can be calculated for every circuit in which the values of the various components are known. When a time equal to four time constants has elapsed after a circuit has been switched on, the transient current flows have dropped virtually to zero, and the steady current is flowing.

The steady current

The amount of steady current flowing in the circuit is obviously affected by the resistance of the circuit. Circuits in which a capacitor is connected in series will have zero steady current because a capacitor acts as an insulator, or open circuit, for d.c. Circuits in which an inductor is connected in series will have a steady value of d.c. current flow that depends on the value of the resistance R of the inductor.

Example. A 2H,1500-ohm inductor has a 1k5 resistor connected in series with it, and is itself connected across a 10V d.c. supply. What steady value of current will eventually flow?

Solution. The total value of resistance is 3k, connected across 10V. The value of steady current flow, in mA, will therefore be 10/3 = 3.3 mA. The two-henry inductance value of the coil is not relevant in this calculation.

Capacitors connected in parallel have no effect on the value of the steady current. Inductors connected in parallel act in the same way as resistors insofar as the steady current is concerned, and have the same effects.

Transient currents

Both inductors and capacitors have a big influence on the transient currents that flow for the instants that elapse just after a d.c. circuit is switched on. The essential facts to remember are:

1   The voltage across a capacitor cannot change instantly, but only at a rate decided by the time constant of the circuit, i.e., its $C \times R$ where C is the value of the capacitance in farads, and R is the value of the resistance in ohms.
2   The amount of current flowing through an inductor cannot change instantly, but only at a rate determined by the time constant of the circuit, i.e., L/R, where L is the value of the inductance in henries and R is the value of the resistance in ohms.

Exercise 6.1

Connect the circuit shown in Figure 6.2(a). The inductor L should be of the high-inductance, low-resistance type. Set the stop-clock to zero, but be ready to start it the instant the circuit is switched on. (An electric clock is obviously ideal for the purpose.)

Note the values of current flow against time elapsed for the first few seconds after switching on, and plot these values on a graph. The shape of the curve you will obtain is shown in Figure 6.2(b).

The results of this exercise prove that when a voltage is applied across an inductor, current flow cannot instantly reach its final value of V/R. The time constant for this circuit can be calculated if the values of L and R are known.

Alternatively, the value of the time constant can be found by inspection of the graph. It is the time needed after switch-on for the current to reach 63% of its final value. After a time equal to four times the duration of the time constant, the current will have practically reached its steady value.

Back-e.m.f. in an inductor

The slow rise of current through a large inductor when a voltage is connected across it is caused by the existence of *back-e.m.f.*, which is a voltage generated by the magnetic field of the inductor acting in the opposite direction to the applied voltage. The back-e.m.f. is particularly large when the rate-of-change of current flow is rapid, as occurs when a circuit containing an inductor is switched off.

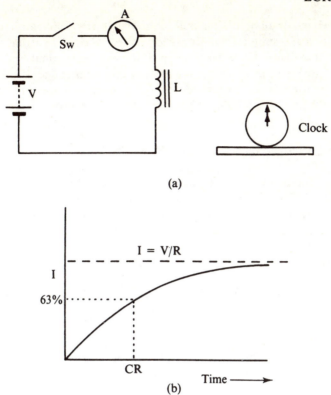

(a)

(b)

Figure 6.2   Circuit for Exercise 6.1

Exercise 6.2

Connect the circuit shown in Figure 6.3. The neon lamp will light only when the voltage across it rises to the order of 80V. Observe the light as the current is switched on and off.

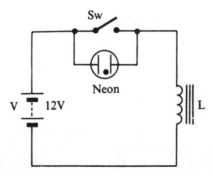

Figure 6.3   Circuit for Exercise 6.2

155

The behaviour of the neon bulb shows that large voltages, considerably greater than the supply voltage, are generated when the current through an inductor is switched off. In many circuits, it is necessary to take precautions to prevent damage to other components from this transient back-e.m.f. voltage. Diodes are for this reason commonly used to protect transistors which switch current in inductive loads, as will be seen in a later chapter.

Charge and discharge in a capacitor

While the voltage across the terminals of a capacitor is increasing, the capacitor is charging (i.e., storing charge). When the voltage across a capacitor is decreasing, the capacitor is discharging, or releasing charge. Both of these processes take time, measured by the time constant $C \times R$ (expressed in seconds).

Exercise 6.3

Connect the circuit of Figure 6.4(a), using a valve or FET voltmeter to measure

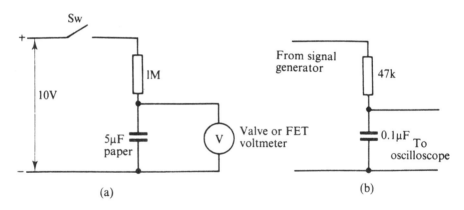

Figure 6.4   Circuits for Exercise 6.3

the voltage across the capacitor. Take a reading of voltage for every second after switch-on, and plot a voltage/time graph. Compare this with the graph developed in Exercise 6.1.

Now connect the circuit of Figure 6.4(b). Set the signal generator to deliver a 1kHz square wave of 2V amplitude peak-to-peak, and observe this trace on the oscilloscope. How does the shape of the trace compare with the shape of the graph line developed in Figure 6.4(a)?

Examine the oscilloscope trace again with input frequencies of 100Hz and 10kHz, using 2V peak-to-peak square waves. Draw the resulting waveforms.

Change the output of the signal generator to sine-wave and repeat the experiment, noting the shape of the output waveforms appearing on the oscilloscope.

The reason for the shape of the graph obtained in Exercise 6.3 is that the voltage across the capacitance plates cannot change instantly when the voltage across the whole circuit is changed.

Another transient effect will be found if a capacitor and a resistor are connected as in Figure 6.5. Again, the voltage between the capacitor plates

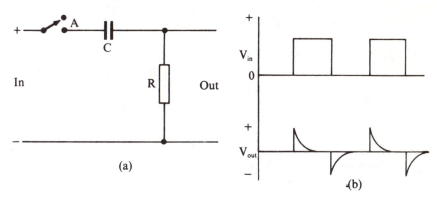

Figure 6.5   Circuits for Exercise 6.3

cannot change instantly, so that when the switch is moved to Position A, voltage at the output at first equals battery voltage (so that the potential between the capacitor plates is zero volts), but then drops as the capacitor charges.

Both circuits of Figure 6.4 are known as *integrating circuits*, that of Figure 6.5 as a *differentiating circuit*. The action of either type of circuit on a square wave is determined by the value of the time constant of the circuit compared to the wave-time (or period) of the square wave.

Summary

A circuit containing capacitors and/or inductors in addition to resistors will, for some time after a steady voltage is applied across it, behave as if the capacitors were open circuits and the inductors small-value resistors.

For a short time after a steady voltage has been either applied or removed, transient currents and voltages will exist in the circuit.

These effects will disappear after a time equal to approximately four time constants, but are very important in wave-shaping circuits.

The time constant is $C \times R$ seconds in a capacitor-resistor circuit and L/R seconds in an inductor-resistor circuit.

Sine-wave-driven circuits

Reactance

Capacitors and inductors in d.c. circuits cause transient current effects only when the applied voltage changes. In an a.c. circuit, of course, the voltage is changing all the time, so that capacitors in such a circuit are continually charging and discharging, and inductors are continually generating back-e.m.f. In a *complex* circuit (as it is called) containing resistors, capacitors and inductors, an alternating voltage will exist across each component which is still proportional to the amount of current flowing through the component, however, so that a form of Ohm's law applies.

In the explanation that follows, the symbols V' and I' are used to mean a.c. values (peak or r.m.s.) of alternating signals. The symbols V and I will have their usual meaning of d.c. values.

In a resistor, $V' = RI'$; and the value of resistance found from the variant form of this equation, $R = V'/I'$, is the same as the d.c. value, V/I.

In a capacitor or an inductor, the ratio V'/I' is called *reactance*, symbolized as X. Because this is a ratio of volts to amperes, the same units (ohms) are used to express its 'resistance', but this resistance is a quantity of a very different kind. A capacitor may, for example, have a reactance of only 1k, but a d.c. resistance that is unmeasurably high. An inductor may have a d.c. resistance of 10 ohms, but a reactance of as much as 5k.

Unlike a resistance, too, the reactance of a capacitor or an inductor is not a constant quantity. A capacitor, for example, has a very high reactance to low-frequency signals and a very low reactance to high-frequency signals.

The reactance of a *capacitor* V'/I' can be calculated from the equation:

$$X_C = \frac{1}{2\pi \times f \times C} \text{ ohms,}$$

where f is the frequency of the signal in Hz and C the capacitance in farads. The tables above show the values of capacitive reactance which are found at various frequencies.

The reactance of an inductor varies in the opposite way, being low for low-frequency signals and high for high-frequency signals. Its value can be calculated from the equation:

$$X_L = 2\pi \times f \times L \text{ ohms,}$$

where f is the frequency in Hz and L is the inductance in henries. The tables that follow show the values of inductive reactance which are found at various frequencies.

Capacitive reactance tables

Audio frequencies Frequency capacitance	20Hz	50Hz	400Hz	1kHz	5kHz	10kHz	20kHz
470 pF	Very high values		847k	339k	68k	34k	17k
2n2	Very high values		181k	72k	15k	7k	3k6
10n	796k	318k	40k	16k	3k	1k6	798R
47n	169k	68k	8k5	3k4	678R	339R	169R
220n	36k	14k	1k8	724R	145R	72R	36R
1µF	8k	3k	400R	160R	32R	16R	8R
10µF	800R	318R	40R	16R	Very small values . . .		
100µF	80R	32R	4R	Small	Very small values . . .		

Radio frequencies Frequency (Hz) capacitance	100k	470k	1M	5M	10M	20M	30M
10pF	160k	34k	16k	3k	1k6	800R	531R
22pF	72k	15k	7k	1k5	724R	362R	241R
100pF	16k	3k4	1k6	318R	159R	80R	53R
470pF	3k4	721R	339R	68R	34R	17R	Small
1n	1k6	339R	160R	32R	16R	Small values	
4n7	339R	72R	34R	—	—	Small values	
22n	72R	15R	7R	—	—	Small values	
47n	34R	7R	—	—	—	Small values	
0.1µF	16R	—	—	—	—	Small values	

Note: Exact values of capacitive reactance have been omitted when calculation gives either very high (near 1M) or very low (a few ohms or less) values, since calculated values are not reliable at these extremes.

All figures have been rounded off, since exact values are never required.

Exercise 6.4

Connect the circuit shown in Figure 6.6. If meters of different ranges have to be used, changes in the values of capacitor and inductor will also be necessary. The signal generator must be capable of supplying enough current to deflect the current meter which is being used.

Connect a 5µF paper capacitor between the terminals, and set the signal generator to a frequency of 100Hz. Adjust the output so that readings of a.c. voltage and current can be made. Find the value of V'/I' at 100Hz.

Repeat the readings at 500Hz and at 1000Hz. Tabulate values of X_C = V'/I' and of frequency f.

Now remove the capacitor and substitute a 0.5H inductor. Find the reactance at 100Hz and 1000Hz as before, and tabulate values of X_L = V'/I' and of frequency f.

Inductive reactance tables

Audio frequencies Frequency inductance	20Hz	50Hz	400Hz	1kHz	5kHz	10kHz	20kHz
20mH	Small values		50R	126R	628R	1k3	2k5
50mH	Small values		126R	314R	1k6	3k1	6k3
100mH	Small	31R	251R	628R	3k1	6k3	12k6
500mH	63R	157R	1k3	3k1	16k	31k	63k
1H	126R	314R	2k5	6k3	31k	63k	126k
10H	1k3	3k1	25k	63k	314k	Large values	
100H	12k6	31k4	251k	—	Very large values		

Radio frequencies Frequency (Hz) inductance	100k	470k	1M	5M	10M	20M	30M
10μH	Small	30R	63R	314R	628R	1k3	1k9
50μH	31R	148R	314R	1k6	3k1	6k3	9k4
200μH	126R	590R	1k3	6k3	12k6	25k	37k7
1mH	628R	2k9	6k3	31k4	62k8	126k	188k
2mH	1k3	6k	12k6	63k	126k	251k	Large
5mH	3k1	14k8	31k4	157k	—	Large values	
10mH	6k3	29k	63k	—	—	Large values	

Note: Exact values have again been omitted when calculation gives either very high or very low values, since calculated values are not reliable at these extremes.
All figures have been rounded off, since exact values are never required.

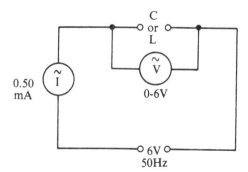

Figure 6.6   Circuit for Exercise 6.4

Next, either remove the core from the inductor (if this is possible) or increase the size of the gap in the core, and repeat the measurements. How has the reactance value been affected by the change?

Phase angle

There is another important difference between a resistance and a reactance, whether it be capacitive or inductive.

With the aid of a double-beam oscilloscope, the waveform of current through a resistor and of voltage across the resistor can be displayed together (see Figure 6.7). This shows, as one might expect, that these waves coincide, with peak

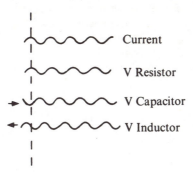

Figure 6.7   Phase shifts

current coinciding with the peak voltage, and so on. But if this experiment is repeated with a capacitor or an inductor in place of the resistor, it will be seen from the figure that the waves of current and voltage do not coincide, but are a quarter-cycle (90°) out of step.

Comparing the positions of the peaks of voltage and of current, it will be seen that:

- with a capacitor, the current wave *leads* (or precedes) the voltage wave by a quarter-cycle; and
- with an inductor, the voltage wave leads the current wave by a quarter-cycle.

Another way of saying the same things is:

- With a capacitor, the voltage wave *lags* (or arrives after) the current wave by a quarter-cycle;
- With an inductor, the current wave lags the voltage wave by a quarter-cycle.

The amount by which the waves are out of step is usually defined as an angle, called the *phase angle*. The reason is that a coil of wire rotating in the field of a magnet generates a sine wave, with one cycle of wave being generated for every turn (360°) of rotation. One half-cycle thus corresponds to 180°, and one quarter-cycle to 90°.

Current and voltage are thus said to be *90° out of phase* in a reactive component such as a capacitor or an inductor.

A convenient way of remembering the order of current and voltage is the word C-I-V-I-L, meaning C − I leads V; V leads I for L. The letters C and L are used to denote capacitance and inductance respectively.

Phasor diagrams

It is only necessary to take a few measurements on circuits containing reactive components to see that the normal circuit laws used for d.c. circuits cannot be applied directly to a.c. circuits.

Consider, for example, a series circuit containing a 10μF capacitor C, a 2H inductor L and a 470-ohm resistor R, as in Figure 6.8. With a 10V voltage V'

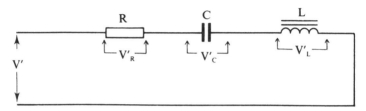

Figure 6.8   Circuit for Exercise 6.5

alternating at 50Hz applied to the circuit, the a.c. voltages across each component can be measured and added together $(V'_C + V'_L + V'_R)$. It will be found that these measured voltages do not add up to the voltage V' across the whole circuit.

Exercise 6.5

Connect the circuit shown in Figure 6.8, with the values given above. Use either a high-resistance a.c. voltmeter or an oscilloscope to measure the voltage V'_R across the resistor and the voltage V'_C across the capacitor. Now measure the total voltage V', and compare it with $V'_R + V'_C$.

Repeat the procedure, substituting the inductor L for the capacitor C and finding V'_L, V'_R and V'. Again compare V' with $V'_R + V'_L$.

The reason why the component voltages in a complex circuit do not add up to the circuit voltage when a.c. flows through it is the phase angle between voltage and current in the reactive component(s). At the peak of the current wave, for example, the voltage wave across the resistor will also be at its peak, but the voltage wave across any reactive component will be at its zero value (see Figure 6.7). Measurements of voltage cannot however, indicate phase angle. They can only give the r.m.s. or peak values for each component, and the fact that these values do not occur at the same time cannot be allowed for by meter measurement. The result is that straight addition of the measured value will inevitably give a wrong result for total voltage, because of the time difference.

Phasor diagrams (often also called *vector diagrams*) are one method of performing the addition so that phase angle is allowed for.

In a phasor diagram, the voltage across a resistor in an a.c. series circuit is represented by the length of a horizontal line drawn to scale. Voltages across reactive components are represented by the lengths of vertical lines, also drawn to scale. If all the lines are drawn from a single point, as in Figure 6.9(a), the

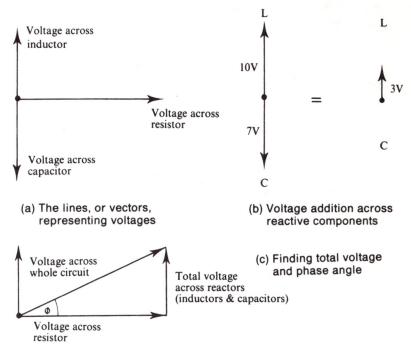

(a) The lines, or vectors,
representing voltages

(b) Voltage addition across
reactive components

(c) Finding total voltage
and phase angle

Figure 6.9   Phasor diagrams for complex series circuits

resulting diagram is a phasor diagram that represents both the phase and the voltage of the wave across each component.

To represent the opposite effects that capacitors and inductors have on the phase, the vertical line representing voltage across an inductor is drawn vertically upwards, and the line representing voltage on a capacitor is drawn vertically downwards.

The phasor diagram can now be used to find the total voltage across the whole circuit. First, the difference between total upward (inductive) and total downward (capacitive) voltage is found, and a line is drawn to represent the size and direction of this difference. For example, if the inductive voltage is 10V and the capacitive voltage 7V, the difference is 3V drawn to scale in the direction of inductive reactance. If the inductive voltage were 10V and the capacitive voltage 12V, the difference would be 2V drawn to scale downwards in the capacitive direction.

The 'net' reactive voltage so drawn is then combined with the voltage across the resistor in the following way. Starting from the point marking the end of the line representing the voltage across the resistor, draw a vertical line, as in Figure 6.9(b), to represent the net reactive voltage in the correct direction, up or down. Then connect the end of this vertical line to the starting point (Figure 6.9(c)).

The length of this sloping line will give the voltage across the whole circuit, and its angle to the horizontal will give the phase angle (ϕ) between voltage and current in the whole circuit.

Impedance

A complex circuit which contains both resistance and reactance possesses another characteristic which is of great importance. This characteristic is known as *impedance*, symbolized by Z. Impedance is measured in ohms, and is equal to the quotient of the values V'/I' for the whole circuit. Its value varies as the frequency of the signal varies.

When impedance is present, the phase angle between current and voltage is neither 90° (as it would be for a reactor) nor 0° (as it would be for a resistor), but some value in between the two. This angle can often be most easily found by using the device of a phasor diagram in a slightly different way, to form what is known as the *impedance triangle*.

In a phasor diagram constructed with this object, separate lines are drawn to represent the resistance R, the reactance X and the impedance Z. In a series circuit, the length of the horizontal line represents the total value of resistance in the circuit, and the vertical line its net value of reactance – upwards as before, for predominantly inductive reactance (Figure 6.10(a)) and downwards for predominantly capacitive reactance (Figure 6.10(b)).

With the values of R and X known and the angle between them a right angle,

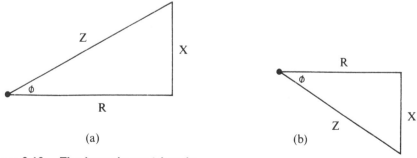

Figure 6.10   The impedance triangle

the Z line can be drawn in, representing the impedance value of the whole circuit. The angle of this line to the horizontal is the phase angle between current and voltage in the circuit.

Another way of working out the relationships between R, X and Z in a complex circuit is to express them by two algebraic formulae:

$$Z = \sqrt{(X_L - X_C)^2 + R^2}$$

and

$$\tan\phi = \frac{X_L - X_C}{R}$$

where Z = total impedance. X_L = inductive reactance. X_C = capacitive reaction, and R = the resistance of the circuit as a whole. A pocket calculator covering a reasonably full range of mathematical functions can now be used to work out the values of circuit impedance and phase angle respectively.

Summary

In an a.c. circuit, capacitors and inductors possess reactance, which is measured in ohms. The amount of reactance depends on the frequency of the a.c. and on the values of the various components.

A 90° phase difference between current and voltage exists in every reactive component.

A circuit which contains resistance as well as reactance possesses also impedance. Impedance, too, is measured in ohms, and has a phase angle somewhere between 0° and 90°. Both phase angle and the value of the impedance itself can be determined either algebraically or by the use of the impedance triangle.

Filters

It is possible to make good use of the way in which reactance varies when frequency is varied.

Figure 6.11(a) shows a single RC circuit. At low frequencies, the reactance of the capacitor will be high, so that there is very little potential-divider action. At higher frequencies, however, the reactance of the capacitor will be less – with the result that the circuit now acts as a potential divider whose output signal voltage is less than its input signal voltage.

Such a circuit is called a *simple low-pass filter*, because it passes low-frequency signals unchanged, but both decreases the amplitude and changes the phase of signals of higher frequency.

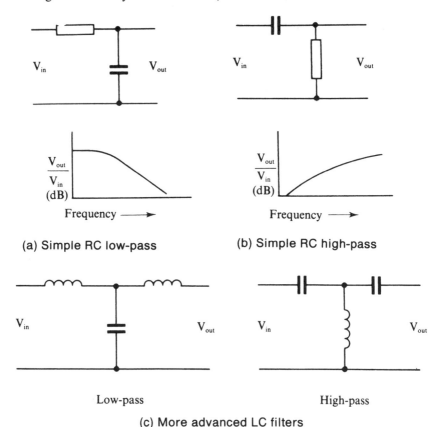

(a) Simple RC low-pass

(b) Simple RC high-pass

Low-pass

High-pass

(c) More advanced LC filters

Figure 6.11   Filters

When the response of a circuit of this type is graphed, the normal method is to show the voltage ratio (attenuation) in decibels and the frequency on a logarithmic scale. For simple circuits using one capacitor and one resistor, the response is 3 dB down at the frequency for which the reactance of the capacitor equals the resistance of the resistor, and from then on the response drops to 7 dB down at double this frequency, and becomes 12 dB down at four times this frequency. From this point on the response drops at 6 dB per octave, meaning 6 dB for each doubling of the frequency. For example, if the response were 3 dB down at 2kHz, it would be 7 dB down at 4kHz, 12 dB at 8kHz, then 18 dB down at 16kHz and so on. The attenuation is sometimes shown plotted upwards, so that the peak attenuation is the highest point on the graph rather than the lowest.

Figure 6.11(b) shows an equally simple *high-pass filter* based on the same principle. At high frequencies, the reactance of the capacitor is so small compared to the resistance value that there is practically no potential-divider action. At low frequencies, the reactance of the capacitor is so high that the signal is attenuated (i.e., reduced in amplitude) and a phase shift takes place.

Figure 6.11(c) shows a couple of more advanced filters whose graphs of amplitude versus frequency have a steeper slope than can be achieved by any simple RC filter.

Resonance

Part of the process of calculating the value of an impedance involves, as you have seen, finding the difference between the values of capacitive reactance, X_C, and inductive reactance, X_L. At low frequencies, X_C is large and X_L small; at high frequencies, X_C is small and X_L large. There must therefore be some frequency at which $X_C = X_L$.

This frequency is called the *resonant frequency*, or the *frequency of resonance*, of the LCR circuit in question. Its symbol is f_r.

A phasor diagram drawn for a series LCR circuit at its resonant frequency will clearly have a zero vertical component of reactance (Figure 6.12). The

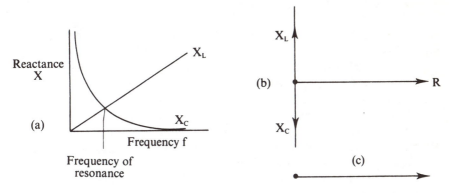

Figure 6.12   Resonance
(a) The resonant frequency of a circuit containing inductance and capacitance is that at which the two reactances are equal
(b) Since the reactance values are always opposite in direction, they cancel out on the phasor diagram
(c) The impedance of a series LCR circuit at resonance is equal to its resistance only

impedance of the circuit will therefore be simply equal to its resistance. The same conclusion can be reached by working out the formula:

$$Z = \sqrt{(X_L - X_C)^2 + R^2}$$

At its resonant frequency, therefore, an LCR circuit behaves as if it contained only resistance, and has zero phase angle between current and voltage.

Exercise 6.6

Connect the circuit shown in Figure 6.13, with component values as follows:

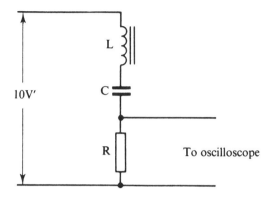

Figure 6.13   Circuit for Exercise 6.6

R = 1k, C = 0.1μF and L = 80mH. The resonant frequency of the circuit is about 1.8kHz.

Set the signal generator to 100Hz, and connect the oscilloscope so as to measure the voltage across the resistor R. This voltage will be proportional to the amount of current flowing through the circuit, because V = R × I.

Now increase the frequency, watching the oscilloscope. The resonant frequency is the frequency at which current flow (and therefore the voltage across R) is a maximum. Note this frequency, and the value of the amplitude of the voltage across R at the resonant frequency.

Measure the voltages across L and across C by connecting the oscilloscope across each in turn. Note the value of these voltages.

Finally, use the oscilloscope to measure the voltage across the whole circuit.

Construct a phasor diagram for the voltages across R, C and L and confirm that this produces an answer for the total voltage. (Remember that the oscilloscope itself will disturb the circuit to some extent, and that the resistance of the inductor has not been taken into account in your calculation.)

A *series-resonant circuit* consists of an inductor, a capacitor and a resistor in series. At the frequency of resonance, the circuit as a whole behaves as if only the resistor were present. Current flow in the circuit will be large if the voltage across the whole circuit remains constant, so that a large voltage exists across each of the reactive components in the circuit.

Generally, however, the voltage across both the capacitor and the inductor will be greater than the voltage across the whole circuit at the frequency of resonance. The ratio: V'_x/V'_z, where V'_x is the voltage across a reactor and V'_z is the voltage across the whole circuit, is called the *circuit magnification factor*, whose symbol is Q. Q can be very large at the frequency of resonance.

The frequency of resonance for a series circuit can be calculated by using the formula:

$$f_r = \frac{1}{2\pi\sqrt{L \times C}}$$

where L is the inductance in henries, C the capacitance in farads, and f the frequency in hertz.

Example. What is the resonant frequency of a circuit containing a 200mH inductor and a 0.05μf capacitor?

Solution. Substitute the data in the equation: $f_r = \dfrac{1}{2\pi\sqrt{L \times C}}$ taking care to reduce both L and C to henries and farads respectively. Thus $L = 200 \times 10^{-3} = 0.2H$, and $C = .05 \times 10^{-6} = 5 \times 10^{-8}F$. Take 2π as being $\simeq 6.3$.

Then

$$f_r = \frac{1}{6.3\sqrt{0.2 \times 5 \times 10^{-8}}}$$

$$= 1587Hz$$

Parallel resonance

A circuit consisting of inductance, capacitance and resistance in parallel will resonate at a frequency given approximately by the same equation:

$$f_r = \frac{1}{2\pi\sqrt{L \times C}}$$

At the frequency of resonance, a parallel resonant circuit behaves like a large value of resistance. Its actual value is $\dfrac{L}{C \times R}$, which is called the *dynamic resistance*. There is no phase angle between voltage and current at the frequency of resonance.

Exercise 6.7

Connect the parallel resonant circuit shown in Figure 6.14 to the series resistor and the signal generator. Find the frequency of resonance, which for the

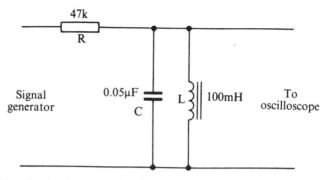

Figure 6.14   Circuit for Exercise 6.7

component values shown will be about 2250Hz, and note that at this resonant frequency, the voltage across the resonant circuit is a maximum.

Now connect another .05μF capacitor in parallel with C, and note the new frequency of resonance.

Remove the additional capacitor, and plot a graph of the voltage across the resonant circuit against frequency, for a range of frequencies centred about the resonant frequency. Observe the shape of the resulting curve, which is called the *resonance or response curve.*

Now add a 10k resistor in parallel with the resonant circuit, and plot another resonance curve, using the same frequency values. What change is there in the shape of the curve?

Repeat the experiment using a 1k resistor in place of the 10k one, and plot all three graphs on the same scale.

The result of these experiments will show that the addition of either capacitance or inductance to a parallel resonant circuit causes the frequency of resonance to become lower. The addition of resistance in parallel has little effect on the frequency of resonance, but a considerable effect on the shape of the resonance curve. The effect of adding a small value of resistance is to lower the peak of the resonance curve – as might be expected because the sum of two resistors in parallel is a net resistance smaller than either. In addition, however, the width of the curve is increased.

A resistor used in this way is called a *damping resistor.* Its effect is to make the resonant circuit respond to a wider range of frequencies, though at a lower amplitude. A damping resistor therefore increases the bandwidth of a resonant circuit, making the circuit less selective of frequency. See Figure 6.15.

When a parallel resonant circuit is used as the load of a tuned amplifier, the tuned frequency is the resonant frequency of the parallel resonant circuit, and the amount of damping resistance employed will determine the bandwidth of the amplifier.

Using LCR circuits allows us to construct bandpass and bandstop filters. The

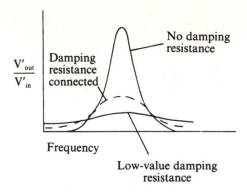

$$\frac{V'_{out}}{V'_{in}}$$

Damping resistance connected

No damping resistance

Frequency

Low-value damping resistance

Figure 6.15   The effect of damping resistors on a parallel resonant circuit

simple LC parallel or series circuits can have resistors added to dampen the resonance so that the resonant effect is reduced but spread over a wider range of frequencies, providing a simple bandpass or bandstop action according to where the resonant circuit is placed. For many purposes, however, these simple circuits do not provide a sharp enough distinction between the pass and the stop bands and much more elaborate filter circuits have to be used. Bandpass and bandstop characteristics are often plotted on a linear scale of frequency so that the bandwidth is easier to read from the graph, but the attenuation is always plotted in terms of decibels.

Calculations on such filters are very difficult, but computer programs can be used to print out a graph of response for any combination of components. Figure 6.16 shows a bandpass filter, using a parallel circuit to determine the bandpass mid-frequency. The graph shows the response of this circuit for the components as follows:

Source and load impedance	75Ω
L_1, L_2	24mH
C_1, C_2	33nF
L_3	6mH
C_3	120nF

Note that the graph produced by the computer program shows that there is some attenuation even at the passband. This is because the computer takes account of the input and output resistances which are usually ignored.

Summary

A circuit containing a capacitor and an inductor will have a resonant frequency (or frequency of resonance) at which the circuit has no phase shift.

A series resonant circuit will have low impedance at the resonant frequency – an impedance equal only to the circuit resistance. A parallel resonant circuit will

171

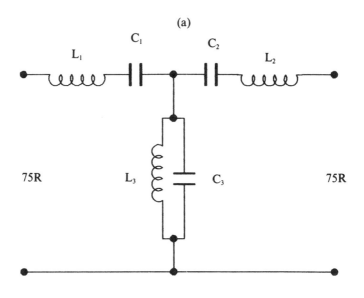

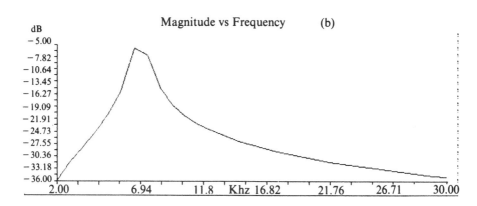

Figure 6.16 **The circuit (a) and the response plot (b) for a bandpass filter. The response has been produced by a computer program**

have a high impedance at resonance (the dynamic resistance), and is used as the load for a tuned amplifier.

The shape of the resonance curve (amplitude plotted against frequency) for a parallel tuned circuit can be altered by adding damping resistors in parallel. This has the effect of lowering the peak of the curve and broadening its base.

Multiple-choice test questions

1   The transient currents in a CR circuit are complete for all practical purposes in a time equal to:
    (a) one time constant
    (b) two time constants
    (c) half a time constant
    (d) Four time constants.

2   A large voltage can be developed from an inductor when:
    (a) there is a small steady voltage across the inductor
    (b) when current through the inductor is switched off
    (c) when a large steady current flows through the inductor
    (d) when a current starts to flow through the inductor

3   A squarewave is fed into two circuits, A and B. A produces brief spikes of pulse waveform, B produces a triangular wave. This means:
    (a) A is a differentiator and B is an integrator
    (b) B is a differentiator and A is an integator
    (c) both circuits are differentiators
    (d) both circuits are integrators.

4   A 1nF capacitor has a reactance of about 3183Ω at 50kHz. The reactance of a 5n capacitor at 20kHz would be approximately:
    (a) 15915Ω (c) 1591Ω
    (b) 636Ω (d) 7957Ω.

5   An inductor and a capacitor are connected together in two different ways and the impedance measured at the resonant frequency. For circuit A the impedance at resonance is 5Ω, for circuit B the impedance is 25k.
    (a) Circuit A contains a 5Ω resistor and circuit B contains a 25k resistor.
    (b) Circuit A is a parallel circuit and circuit B is a series circuit.
    (c) Circuit A has the inductor connected the other way round as compared to circuit B.
    (d) Circuit A is a series circuit and circuit B is a parallel circuit.

6   An a.c. circuit contains a resistor, a capacitor and an inductor. The a.c. voltages across each component are measured, but the sum of the voltages is not equal to the voltage across the whole circuit. This is because:
    (a) such voltages never add up correctly
    (b) the phase angles are not being taken into account
    (c) voltmeters do not read correctly on a.c.
    (d) only the voltage across the resistor should be counted.

7 Oscillators and waveform generating

Summary

Positive feedback, oscillation. Sine wave oscillators. Crystal control. RC oscillators. Aperiodic oscillators – the multivibrator. Bistable and monostable. Scale-of-two counter. Astable. Frequency control. Wave-shaping. Timebase generators. The phase-locked loop. Schmitt trigger. Miller integrator. Unijunction oscillators.

Positive feedback and oscillation

The effects of negative feedback of a.c. signals on a voltage amplifier were described in Chapter 4. It will be recalled that negative feedback is achieved by subtracting a fraction of the output signal from the input signal of an amplifier. In practice, this is done by adding back the feedback signal in antiphase, so that feedback from an output which is in antiphase to an input is always negative unless some change of phase occurs in the circuit used to connect the output to the input (see later, under RC oscillators).

If a signal which is in phase with the input is fed back, the feedback becomes positive. Positive feedback takes place when a fraction of the output signal is added to the signal at the input of an amplifier, so increasing the amplitude of the input signal. The result of positive feedback is higher gain (though at the cost of more noise and distortion) if the amount of feedback is small. If the amount of feedback is large, the result is oscillation.

An amplifier oscillates when:

- the feedback is positive at some frequency; and

174

- the voltage gain of the amplifier is greater than the attenuation of the feedback loop (see Chapter 4 again to recall the concept of loop gain).

If, for example, 1/50th of the output signal of an amplifier is fed back in phase, oscillation will take place if the gain (without feedback) of the amplifier is more than 50 times, and if the feedback is still in phase.

Oscillator feedback circuits are arranged so that only one frequency of oscillation is obtained. This can be done by ensuring either:

- that the feedback is in phase at only one frequency; or
- that amplifier gain exceeds feedback loop attenuation at one frequency only; or
- that the amplifier switches off entirely between conducting periods.

Oscillator circuits are of two types. *Sine-wave oscillators* use the first two methods above for ensuring constant frequency operation. *Aperiodic* (or *untuned*) oscillators, such as multivibrators, make use of the third method. Oscillators are thus equivalent to amplifiers which provide their own inputs. They also convert into a.c. the d.c. energy from the power supply.

Most oscillator circuits operate in Class C conditions. Even if the transistors start off with some bias current flowing, the action of the oscillator will turn off the bias for quite a large proportion of the complete waveform, making the transistor once it is oscillating operate in Class C. The reason for this will be clearer as you read through this chapter; it arises because a tuned circuit connected to an oscillator will continue to oscillate even when the transistor is no longer conducting.

Sine-wave oscillators

A sine-wave oscillator consists of an amplifier, a positive feedback loop, and a tuned circuit which ensures that oscillation occurs at a single definite frequency. In addition there must be some method of stabilizing the amplitude of the oscillations so that the oscillation neither stops, nor builds up to such an amplitude that the wave becomes distorted by reason of bottoming or cut-off.

The most common types of sine-wave oscillator are those which operate at radio frequencies, such as are used in the local oscillators for superhet receivers which use LC tuned circuits to determine the oscillating frequency.

The Hartley oscillator

Like most oscillator circuits, the Hartley oscillator exists in several forms; but the circuit of Figure 7.1 is a much-used type. The tuned circuit $L_1 C_2$ has its coil tapped to feed a fraction of the output signal back through C_4 to the emitter of Tr1. Since

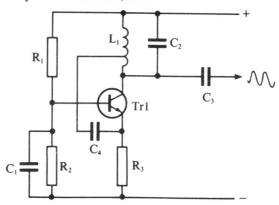

Figure 7.1   One form of the Hartley oscillator

an output at the emitter is always in phase with the output at the collector, this feedback signal is positive. The base voltage of Tr1 is fixed by the values of the resistors R_1 and R_2 with C_1 acting as an a.c. by-pass capacitor.

The amplitude of the oscillation is limited at the emitter, because the transistor will cut off if emitter voltage rises to a value more than about 0.5V below base voltage. The distortion of the wave-shape caused by this limiting effect is smoothed out by the 'flywheel' effect of the tuned (or 'tank') circuit L_1 C_2, which produces a sine-wave voltage at the resonant frequency even when the current waveform is not a perfect sine-wave.

Irrespective of how its feedback is arranged, the Hartley oscillator can always be recognized by its use of a tapped coil. Its frequency of oscillation, as is the case with all oscillators using LC tuned circuits, is given by the formula:

$$f_o = \frac{1}{2\pi\sqrt{LC}}$$

Example. What is the oscillating frequency of a Hartley oscillator which has a 15µH inductor and a 680pF capacitor in its tuned circuit?
Solution. Substitute the data in the equation:

$$f_o = \frac{1}{2\pi\sqrt{LC}}$$

Then

$$f_o = \frac{1}{2\pi\sqrt{15 \times 10^{-6} \times 680 \times 10^{-12}}}$$

$$= \frac{1}{2\pi\sqrt{10200 \times 10^{-18}}}$$

$$= \frac{1}{2\pi \times 1.009 \times 10^{-7}}$$

$$= 1.58 \times 10^6 \text{ Hz, or 1.6MHz approx.}$$

Faults which can cause failure in an oscillator of this type include the following:

- bias failure caused by breakdown of either R_1, R_2, or R_3;
- a faulty by-pass capacitor;
- a leaky or o/c coupling capacitor C_4; or
- faults in either C_2 or L_1.

C_2 should be of the silver-mica type of capacitor. Some ceramic capacitors will not permit oscillation because of 'lossy action' – which means that the capacitor dissipates too much power to permit the circuit to resonate properly.

The Colpitts oscillator

The example of this circuit shown in Figure 7.2 demonstrates its basic similarity to

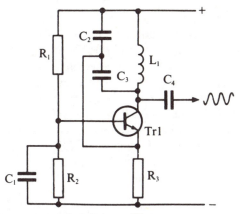

Figure 7.2   One form of the Colpitts oscillator

the Hartley oscillator. Instead of using a tapped coil, however, the Colpitts oscillator employs the combination of C_2 and C_3 to tap off a fraction of the output voltage to feed back into the base. This latter is biased and by-passed in the same way as in the Hartley circuit.

The same remarks about circuit operation, and about the several possible circuit configurations apply to the Colpitts as to the Hartley oscillator, but the formula for determining the frequency of oscillation is slightly different. Because the capacitors C_2 and C_3 are in series, it is the series combination C' (in which $1/C' = 1/C_1 + 1/C_2$) which tunes L_1 to give the output frequency. The formula therefore becomes:

$$f_o = \frac{1}{2\pi\sqrt{L_1 C'}}$$

Example. What is the oscillating frequency of a Colpitts oscillator using an inductor of 25µH, and capacitors of 470pF and 4700pF in series?

177

Solution. The series capacitors have a total capacitance value, C'. (in pF) such that I/C' = 1/470 + 1/4700. Therefore C' = 427pF.

Substituting these data in the formula, one gets:

$$f_o = \frac{1}{2\pi\sqrt{25 \times 10^{-6} \times 427 \times 10^{-12}}}$$

$$= \frac{1}{2\pi\sqrt{10675 \times 10^{-18}}}$$

$$= \frac{1}{2\pi \times 1.03319 \times 10^{-7}}$$

$$= 1.54 \times 10^7 \text{Hz. , or } 15.4\text{MHz}$$

Exercise 7.1

Construct the Colpitts oscillator shown in Figure 7.2 with the following component values: $C_1 = 0.1\mu F$: $C_2 = 0.01\mu F$: $R_1 = 10k$: $R_2 = 1k5$: $R_3 = 1k$: $C_3 = 0.001\mu F$. The inductor, L_1, should consist of 50 turns of 28-gauge enamelled copper wire close wound on a 10mm diameter former fitted with a ferrite core. (Alternatively, the coil can be wound directly on a ferrite rod of the same diameter.)

Check the circuit and connect to a 12V power supply. Connect the collector of the transistor to the Y-input of an oscilloscope, and link the negative line to the oscilloscope earth. Switch on, and adjust the oscilloscope to show the waveform from the oscillator. Measure the amplitude and frequency of the output wave.

Now observe the effects on the amplitude and frequency of oscillation of the following changes:

(a) Increasing the supply voltage to 15V.
(b) Connecting an additional $0.001\mu F$ capacitor in parallel with L_1.
(c) Reducing bias by connecting a 2k2 resistor in parallel with R_2.
(d) Increasing bias by connecting a 22k resistor in parallel with R_1.
(e) Reducing feedback by connecting a $0.1\mu F$ capacitor in parallel with C_2.
(f) Increasing feedback by replacing C_2 by a 470pF capacitor.
(g) Reduce value of R_3 to 100R.

Note that changes (e) and (f) will both inevitably affect the frequency (if oscillation continues) because the capacitance of the tuned circuit has been changed in value.

Note that the oscilloscope is the most certain way of checking that the oscillator is working, and also of measuring the frequency. Another check is to measure the current drawn by an oscillating circuit, which will be much larger than normal if the circuit is not oscillating (because the oscillator operates in Class C when oscillating, requiring less bias current).

Tuned-load oscillators

Oscillators of this type have a tuned circuit as the load of the transistor and use another component for feedback. The tuned-collector feedback oscillator shown in Figure 7.3 uses a feedback winding placed physically close to the winding of L_1 to

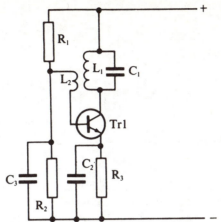

Figure 7.3   A tuned-collector oscillator circuit

extract a fraction of the output signal, to invert its phase, and to feed it back to the base.

Remember that the output of a single-transistor common-emitter amplifier is always phase-inverted, so that another inversion must be carried out if feedback is to be taken from the collector to the base.

Amplitude limitation is carried out in this circuit by the bottoming and cut-off action, and the output is smoothed into a sine-wave by the resonant oscillations of the tuned circuit.

Failure of the oscillator circuit shown in Figure 7.3 can be caused by the by-pass capacitor C_2 going o/c, as well as by any of the biasing faults which can cause the Hartley oscillator to fail.

Another type of tuned-load oscillator is shown in Figure 7.4. This circuit uses a capacitor of very small value, C_3, to feed back part of the signal from the collector to the emitter. This type of oscillator is commonly used at high frequencies as a local oscillator in TV or FM receivers.

Crystal oscillators

The use of a quartz crystal in place of a tuned circuit in an oscillator gives much greater stability of frequency than can be achieved in the same conditions by any LC circuit.

179

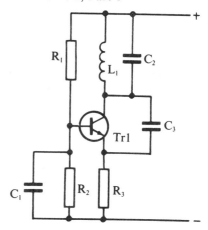

Figure 7.4   An alternative feedback oscillator

There exists a great variety of crystal oscillator circuits, some of which use the crystal as if it were a series LC circuit, with others using it as part of a parallel LC circuit in which the crystal replaces the inductor.

The example of a crystal oscillator shown in Figure 7.5 is a form of Colpitts

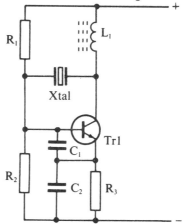

Figure 7.5   A crystal oscillator circuit

oscillator, but with the crystal providing the frequency-determining feedback path and some of the 180° phase shift between collector and base. The choke L_1 acts as the collector load across which an output signal is developed and also contributes to the phase shift.

In normal use, crystal oscillators are extremely reliable, but excessive signal current flowing through the crystal can cause it to break down and fail. The usual comments concerning bias and decoupling components apply here also.

Note that all the circuits for sine-wave oscillators can be constructed with

MOSFETs in place of bipolar transistors. The use of MOSFETs has several advantages, particularly where crystal oscillators are concerned, because no input current is required at the gate of a MOSFET.

RC oscillators

Oscillators required to operate at low frequencies cannot use LC tuned circuits because of the large size of inductor that would be needed. An alternative construction is the RC oscillator, of which Figure 7.6 shows the basic outline of one type –

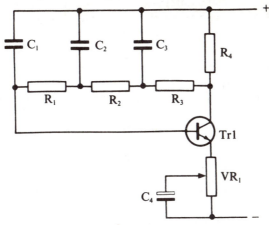

Figure 7.6   A phase-shift oscillator

known as the *phase-shift oscillator*.

Every RC potential divider must attenuate and shift the phase of the signal at the collector of the transistor. If the total phase shift at a given frequency is 180°, the signal fed back to the base will be in phase with the signal at the collector, and the circuit will oscillate.

The output waveform will not, however, be a sine-wave unless the gain of the amplifier can be so controlled that it is only just enough to sustain the oscillation. All RC oscillators therefore require an amplitude-stabilizing circuit, which is usually provided by a negative feedback network. This network usually includes a component such as a thermistor whose resistance decreases as the voltage across it increases. In this way, an increase of signal amplitude causes an increase in the amount of negative feedback – which in turn causes the amplifier gain to decrease, so correcting the amplitude of oscillation.

Exercise 7.2

Construct the phase-shift oscillator of Figure 7.6, using the following values:

R_1, R_2, R_3 220k
R_4 4k7
VR_1 1k
C_1, C_2, C_3 1nF
C_4 100μF 15V
Tr1 2N3056, BC107, BFY50 or similar general-purpose NPN transistors.

Adjust the potentiometer so that the circuit is just oscillating as detected by an oscilloscope connected to the collector of the transistor. Note that low-gain transistors may require the use of lower values for R_1, R_2, R_3.

With the circuit oscillating, note the d.c. voltage levels at the emitter, base and collector of Tr1. Sketch two cycles of the output waveform. Note the peak-to-peak output voltage, and periodic time of the output, and calculate the frequency. Note how the waveshape changes as VR_1 is adjusted.

Repeat the d.c. voltage measurement when the circuit is not oscillating (adjust VR1 to stop oscillation).

What is the effect of replacing one of the resistors R_1, R_2, or R_3 with a 47Ω value?

Another type of RC oscillator is shown in outline in Figure 7.7, in which only the

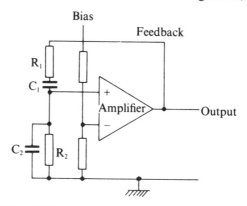

Figure 7.7   The Wien Bridge

components which determine frequency are labelled. This oscillator is known as the *Wien Bridge*. It will be noted that the circuit uses feedback to the non-inverting (in-phase) input of the amplifier.

These RC circuits, also, require amplitude-stabilization if they are to produce sine-waves of good quality, but they are capable of providing very low distortion figures (of the order of only 0.01 %) by good design. An example of a Wien Bridge oscillator circuit with provision for amplitude stabilization is shown in Figure 7.8. The amplifier itself is a 741 IC.

A distorted output from a RC oscillator is nearly always caused by failure of a component in the amplitude-stabilizing circuit. Lack of output is generally due to a

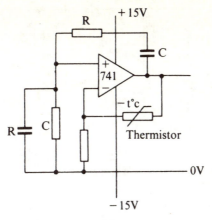

Figure 7.8   A Wien Bridge oscillator with amplitude stabilization

sudden loss of gain, such as could be caused by bias failure or by the failure of a decoupling capacitor.

Summary

Small amounts of positive feedback applied to an amplifier increase the gain of the amplifier. Larger amounts of positive feedback cause oscillation.

An oscillator requires three things: an amplifier possessing sufficient positive feedback; a circuit sensitive to frequency; some method of limiting output amplitude.

Sine-wave oscillators at radio frequencies can use LC tuned circuits or crystals to control the frequency of the sine-wave. Phase-shift or other RC circuits are used to generate sine-waves at audio and other low frequencies.

The shape of the waveform will always be poor unless automatic amplitude-stabilization circuits are employed in oscillator design.

Multivibrators

Multivibrators comprise a family of three circuits, only one of which is in a true sense an oscillator. Each of the families consists of two amplifier stages having large amounts of positive feedback, so that the transistors are rapidly switched from the bottomed to the cut-off states and back again successively. The output waveforms produced by such rapid alternate switching are steep-sided square waves.

The basic configuration is shown in Figure 7.9. At switch-on, one or other of the two transistors starts to conduct faster than does its fellow, by reason of minor differences in their circuits. Say this faster conductor is Tr1. Tr1 collector voltage will now fall faster than will that of Tr2. The fall is passed to Tr2 base as a negative-going signal serving to reduce its forward bias, and so its collector current. The

183

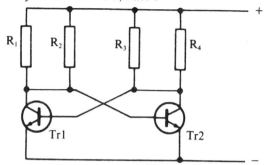

Figure 7.9   The basic multivibrator circuit

collector voltage of Tr2 rises and is passed to Tr1 base as a positive-going signal, so that its conduction increases even further. This constantly-renewed regenerative action rapidly results in Tr1 becoming saturated (or *bottomed*) and Tr2 being cut-off.

The circuit remains in this stable state until it is forced to change.

The bistable multivibrator

A bistable is a multivibrator circuit which employs direct coupling throughout, and which remains in either of two states until it is triggered by a pulse which switches it over to the other state.

Figure 7.10 shows a typical bistable circuit using two transistors. In one stable

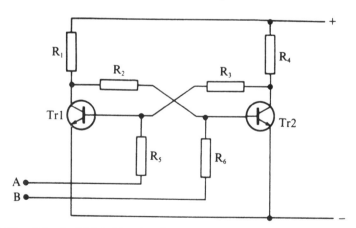

Figure 7.10   A typical bistable circuit

state, Tr1 is conducting with Tr2 cut off. In the other state, the conditions are reversed. Both states are stable.

Trigger pulses

Triggering can be achieved either by applying a negative pulse to the base of the transistor which is conducting or a positive pulse to the base of the transistor which is cut off. The arrangement shown in Figure 7.11 uses *steering diodes* which cause a

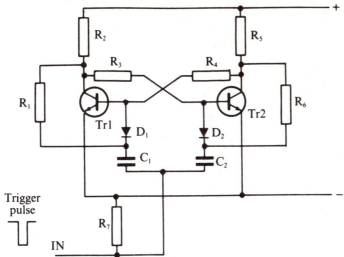

Figure 7.11   A bistable scale-of-two counter with steering diodes

negative pulse to be automatically steered to the base of the transistor which is conducting, so that the circuit switches over. The action is as follows.

Say that Tr1 is conducting, with its base at about 0.5V and its collector voltage bottomed at about 0.2V. The connection of R_1 to the cathode of D_1 ensures that this point is at a low voltage also – with the result that D_1 can conduct if a negative pulse of low amplitude is received through C_1. At the same time, Tr2 is cut off, with its collector voltage high at about supply voltage. The connection of R_6 to the cathode of D_2 biases this diode in the reverse direction, with about 6V on the cathode (given a 6V supply line) and about 0.2V on the anode.

When a negative pulse of small amplitude (between 1V and 6V peak-to-peak) is injected into the circuit at the input, the only possible conduction path is through D_1. Tr1 is therefore cut off, so switching the circuit. As usual, the positive feedback causes the change-over to be very rapid, being completed in something of the order of a microsecond or less.

The change-over also has the effect of switching over the bias voltages on the diodes, so that the next trigger pulse passes easily through D_2 and switches the circuit back to its original state.

Two complete trigger pulses are needed to cause the collector voltage of either Tr1 or Tr2 to go through the complete cycle, so that the circuit gives one complete pulse OUT for two trigger pulses IN. For this reason, it is often known as a *scale-*

of-two, or *binary counter.* The waveforms are shown in Figure 7.12.

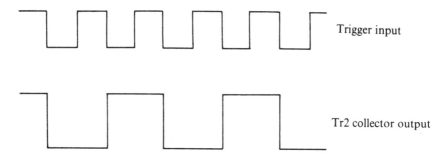
Trigger input

Tr2 collector output

Figure 7.12   Bistable counter waveforms

Exercise 7.3

Construct the bistable multivibrator circuit of Figure 7.10 using the following component values:

R$_1$, R$_4$ 4k7
R$_2$, R$_3$ 47k
R$_5$, R$_6$ 100k
Tr1, Tr2. Any general-purpose NPN transistors such as BC107, BFY50.

Check the circuit, then apply a 9V supply. Use a voltmeter to show that one collector will be at a low voltage and the other at a high (nearly 9V) voltage. Show that the circuit will change over when one of the terminals, A or B, is very briefly connected to +9V. The terminal that has to be used is the one for the transistor that is cut off (high collector voltage). The alternative method of switching is to connect the terminal of the conducting transistor briefly to a negative voltage of sufficient amplitude (− 20V in this case) to cut off the conducting transistor. How can the value of a suitable voltage be calculated?

Now rearrange the circuit so as to correspond to that of Figure 7.11, using the following additional component values:

R$_1$, R$_6$ 100k
C$_1$, C$_2$ 10nF
D$_1$, D$_2$ 1N4001 or any general-purpose diode
R$_7$ 100k

Show that the circuit will remain stable with one transistor conducting and the other shut off until a negative pulse of sufficient amplitude is input. This pulse does not need to have a large amplitude because it is delivered through a diode directly to the base of a transistor; an amplitude of 1–2V should be sufficient. The pulse can be obtained from a single-pulse generator, or from a switch.

The monostable multivibrator

The monostable is a triggered multivibrator circuit which uses one capacitor coupling and one direct coupling in the positive feedback loop between the two transistors. An example is shown in Figure 7.13.

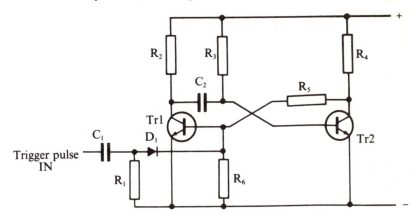

Figure 7.13   The monostable

In this circuit, current flowing through R_3 into the base of Tr2 keeps the transistor switched on, with its collector voltage consequently bottomed. Because the collector voltage of Tr2 is very low, the base voltage of Tr1, which is fed through the potential divider R_5–R_6, will be very low also. The circuit will remain in this state until it is triggered by a positive-going pulse at the input.

In the waveform sketches of the Figure 7.13 monostable given in Figure 7.14, the trigger pulse causes Tr1 to conduct and the positive feedback loop acts to cause the usual switchover action, with Tr1 now conducting and Tr2 cut off. C_2 then charges through R_3; and when the voltage at the base of Tr2 reaches about 0.5V, Tr2 starts to conduct and the positive feedback ensures that the circuit snaps back to its original state.

C_1-R_1 forms a differentiating network to ensure a sharp edge to the trigger pulse. D_1 cuts off the negative-going pulse which C_1-R_1 produces, leaving only a positive trigger pulse.

The base of Tr1 is isolated from any negative pulses (which would shorten the timed period) by the diode D_1. The output pulse at the collector of Tr2 is a positive pulse having a duration of about $0.7C_2R_4$ (with C in farads and R in ohms).

Example. What is the output pulse time of a monostable with a 0.05µF coupling

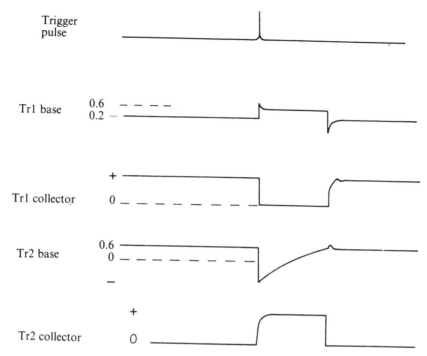

Trigger pulse

Tr1 base $\quad$ 0.6 / 0.2

Tr1 collector $\quad$ + / 0

Tr2 base $\quad$ 0.6 / 0 / −

Tr2 collector $\quad$ + / 0

Figure 7.14 .Monostable waveforms

capacitor and a 68k bias resistor?

Solution. Pulse time = 0.7CR

$$= 0.7 \times .05 \times 10^{-6} \times 68 \times 10^{3}$$
$$= 2.38 \times 10^{-3}$$
$$= 2.4\text{ms (approx.)}$$

The astable multivibrator

The astable multivibrator is a two-transistor circuit which oscillates continuously, producing a wave of approximately square shape at the collector of each transistor. The most common form of the circuit is shown in Figure 7.15, in which the waveforms produced are also shown on the same time scale, and with the same starting times, so that the action of the circuit can be more easily followed.

Say that, as soon as the circuit is switched on, Tr1 conducts. The current flowing through R_1 will keep the collector voltage of Tr1 low, so that the base voltage of Tr2 is also held low because of the coupling capacitor C_1. C_1 now starts to charge, because of current flowing through R_2.

When the voltage at the base of Tr2 reaches about 0.5V, this transistor also starts to conduct. There is a positive feedback loop round Tr1 → C_1 → Tr2 → C_2, so that

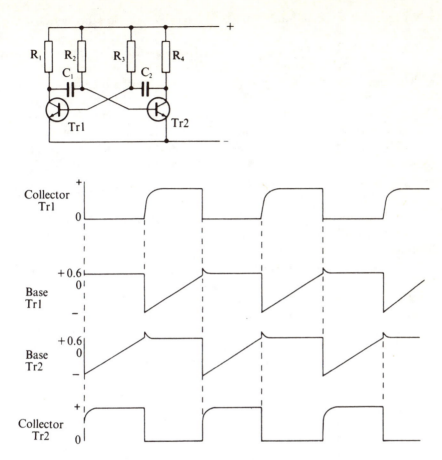

Figure 7.15   The basic multivibrator circuit, and waveforms

once Tr2 starts to conduct, the current increases very quickly until the collector voltage of Tr2 bottoms, when Tr2 ceases to amplify.

The sudden drop in voltage (almost 6V in the example given) at the collector of Tr2 will be transmitted by C_2 to the base of Tr1. This base was formerly at a voltage of about 0.5V, so a 6V drop will cause the base voltage to reach about -5.5V, cutting off Tr1 completely. C_2 now charges because of current flowing through R_3, until the voltage at the base of Tr1 reaches 0.5V. At that point Tr1 starts to conduct again and the positive feedback again causes another rapid switchover – this time to the state where Tr1 is bottomed and Tr2 cut off.

This swiftly-changing and self-perpetuating cycle continues automatically until power is switched off.

The important features of the flip-flop circuit are:

189

- each transistor spends most of the duration of every cycle either bottomed or cut off.
- The 100% positive feedback causes very rapid switching, so that the output waveforms have steep sides.
- The time between switchings is decided by the charging times of C_1 and C_2 through R_2 and R_3 respectively. The duration of a complete cycle is approximately $0.7\,(C_1R_2 + C_2R_3)$, with the time expressed in μs if C is given in microfarads and in seconds if C is given in farads. R is always expressed in ohms.
- The frequency of the output is easily changed by making small changes in either the timing components C_1, R_2, C_2, R_3 or in the bias of the transistors.

Example. What is the approximate frequency of a multivibrator fitted with 0.001μF coupling capacitors and 56k bias resistors?

Solution. The duration of one cycle is approximately $0.7(C_1R_2 + C_1R_3)$, with $C_1 = C_2 = 0.001$μF and $R_2 = R_3 = 56$k. Since in this example $C_1R_2 = C_2R_3$, it follows that $T = 2 \times 0.7 \times CR$.

Substituting, the formula becomes: $2 \times 0.7 \times 0.001 \times 10^{-6} \times 56 \times 10^3$ seconds, and $T = 7.84 \times 10^{-5}$ seconds, or 784 milliseconds.

Since $f = 1/T, f = 1.27 \times 10^4$Hz, or 12.7kHz.

The ease with which the frequency of the astable multivibrator can be varied is put to use in two ways. The first is to produce a circuit such as is shown in Figure 7.16, in which circuit frequency can be varied at will by adjustment of the

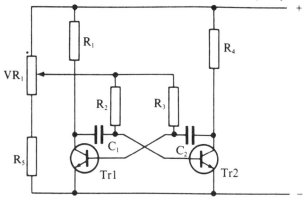

Figure 7.16   A multivibrator circuit with frequency control

component VR_1.

The second is to synchronize the oscillator to trigger pulses. A trigger pulse applied to one base causes the switchover to take place at the very moment of arrival of the pulse. The result is that the frequency of the astable will change to the frequency of the trigger pulse – provided that the frequency of this trigger pulse is higher than the free-running frequency of the astable circuit itself. A typical such circuit is shown, with its waveforms, in Figure 7.17

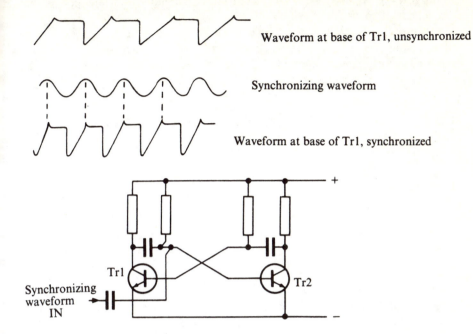

Waveform at base of Tr1, unsynchronized

Synchronizing waveform

Waveform at base of Tr1, synchronized

Figure 7.17   A synchronized multivibrator

The use of resistors and capacitors for timing an astable multivibrator can be replaced by using one long time-constant coupling and a crystal, with the crystal then controlling the timing. Much better results are obtained using digital ICs, however, so that crystal-controlled astable oscillators are not generally constructed using discrete transistors.

Exercise 7.4

Construct an astable multivibrator, using the circuit of Figure 7.15 and the following component values:

R_1, R_4 4k7
R_2, R_3 47k
C_1, C_2 10nF
Tr1, Tr2. Any general purpose NPN transistors such as BC107, BFY50 Power supply voltage 9V.

Check your circuit and then apply a 9V power supply. Use an oscilloscope to view the waveform at one collector and at the base of the same transistor. If a double-beam oscilloscope is available, view both waveforms together. Sketch the waveforms that are obtained. Note the peak-to-peak amplitude and time period, and calculate the frequency.

Measure the d.c. voltage levels at either collector and at either base and note these readings. Find the effect on waveform and frequency when one of the coupling capacitors is changed to 47nF (leaving the other unchanged).

Exercise 7.5

Construct a monostable multivibrator using the circuit of Figure 7.13 and the following values:

R_2, R_4 4k7
R_3, R_5 47k
R_1, R_6 100k
C_1 10nF
C_2 1nF
Tr1, Tr2. Any general-purpose NPN transistors such as BC107, BFY50 Supply voltage +9V.

Connect the input of the circuit to a signal generator which is set to produce square waves of 500Hz. Use an oscilloscope to examine the output – note that the amplitude of the input needs to be sufficiently high to be able to trigger the circuit so as to obtain an output.

When the circuit is producing pulses at the output, measure the peak-to-peak amplitude, and find the repetition rate. Measure the d.c. voltage levels at the emitter, base and collector of each transistor. Repeat the measurements when no input waveform is present.

Summary

Multivibrators are aperiodic oscillators, which means that they contain no tuned circuits.

In the usual two-transistor circuit, the large amount of positive feedback ensures that both transistors conduct for only a very short time. The normal state is to have one transistor conducting fully and the other cut off.

The astable multivibrator oscillates continuously; the monostable gives an output pulse of pre-determined duration when a trigger pulse is injected. In both types, the timing is determined by the time constant of a CR circuit or circuits.

The bistable multivibrator changes from one stable state to another every time a trigger pulse appears at the input.

Wave-shaping circuits

Wave-shaping circuits are circuits designed to alter the shape of waves other than sine-waves. The two most important such circuits are the *differentiating circuit* and the *integrating circuit*. Either can be constructed, in a simple form by the appropriate connection of a single capacitor and a single resistor.

Figure 7.18 shows a differentiating circuit. A sudden rise of voltage at the input,

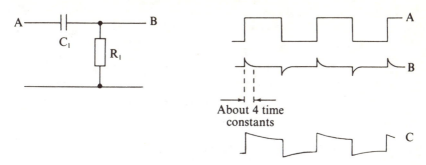

About 4 time
constants

Figure 7.18   The differentiating circuit, and waveforms

Point A, will cause an equal rise of voltage at the output, Point B, because the voltage on one plate of the capacitor will momentarily follow the voltage on the other. The capacitor will now charge, however, through R_1, until Point B reaches zero volts. This takes a time of approximately four times the time-constant of the circuit, R_1C_1 (with R in ohms, C in farads and time in seconds).

If the voltage at Point A should now suddenly drop, the output voltage will quickly drop by the same amount, and then more slowly return to zero as the capacitor discharges. The time required for the operation is again about four time-constants.

The differentiating circuit (as will be seen from waveform B) produces sharp pulses, alternatively positive and negative-going, from inputs featuring sharp rises and falls of voltage. Its only effect on a sine-wave is to attenuate the wave and bring about a shift of phase.

Note that if the time-constant is made too long, the circuit behaves rather like a coupling network, so that the output becomes that of waveform C.

An integrating circuit is shown in Figure 7.19. A sharp rise of voltage at the input, point A causes current to flow through the resistor R to charge the capacitor C. The charging will be complete in about four time-constants; but if the voltage is suddenly reduced before charging is complete, the capacitor will discharge, on the same time-constant, through R. An integrating circuit produces slowly rising or falling waveforms from sharp voltage changes, its action being thus the direct opposite of that of the differentiating circuit.

Note that a unidirectional pulse passing into an integrating circuit will produce a unidirectional output. The output of the differentiating circuit for the same input pulse is a pair of pulses whose average voltage value is zero.

The use of a differentiating circuit has already been illustrated in the timing action of the astable and monostable multivibrators. The names 'differentiating'

193

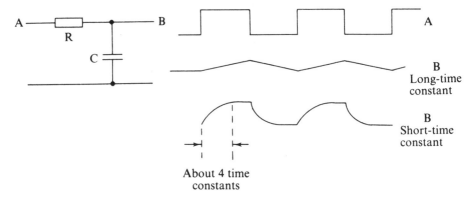

Figure 7.19   The integrating circuit, and waveforms

and 'integrating' are taken from mathematical operations which change equations in a similar way.

The sawtooth generator

The ability of the integrator to convert a square wave into a wave with sloping sides makes it the natural basis for circuits used to generate the sawtooth waveform which produces *timebases*.

A simple timebase circuit is shown in Figure 7.20. While Tr1 is conducting, the voltage at its collector is very low and C_1 is discharged. A negative-going pulse arriving (from an asymmetric astable multivibrator, for example) at the base of Tr1 will cut off Tr1, and C_1 starts to charge through R_2. This integrating action generates a slow-rising waveform which forms the sweep part of the sawtooth.

As the trailing edge of the pulse reaches its base, Tr1 is switched on again. C_1 rapidly discharges through Tr1 and causes the rapid flyback at the end of the sweep waveform.

Tr2 is an emitter-follower acting as a buffer stage to prevent the waveform across C_1 from being affected by the input resistance of circuits to which the sawtooth is coupled.

In such a simple circuit, the waveform across the capacitor will in practice be approximately a straight line if C_1 is only allowed to charge to a small fraction of the supply voltage. Thereafter, the waveform will tend to bend over towards the horizontal. To prevent this, it is important to make the time-constant $R_1 C_2$ much longer than the duration of the square pulse applied to the base of Tr1.

Sawtooth-generator circuits normally use considerably more elaborate circuits than that shown in Figure 7.20, with the object of ensuring that the sweep voltage remains linear. One type of sawtooth generator uses constant-current circuits to

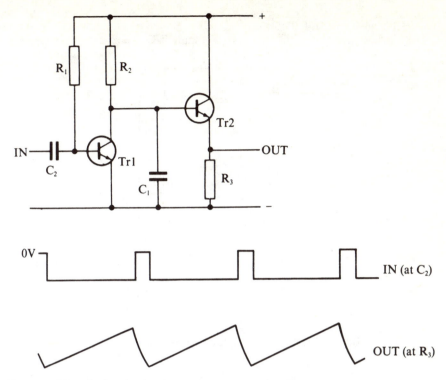

Figure 7.20 A simple timebase, or sweep, circuit

replace R_2. Another, the Miller timebase, uses negative feedback to keep the sweep waveform truly linear.

Summary

Wave-shaping circuits do not change the shape of sine-waves, but affect all other waveforms.

Differentiating circuits form sharp pulses of voltage from the steep edges of square waveforms. Integrating circuits change steeply rising or falling voltages into smoothly changing waveforms.

The very important sawtooth waveform is generated by the action of an integrating circuit on a square-wave input.

The phase-locked loop

The phase-locked loop (PLL) is a circuit which has been much more widely used since it became available in the form of an IC (e.g. the National Semiconductor Co.'s NE567). The information which follows refers to the IC form of the PLL only, since the circuit is nowadays seldom encountered in discrete form save when high frequencies are being handled.

The block diagram of a PLL is shown in Figure 7.21. The heart of the circuit is

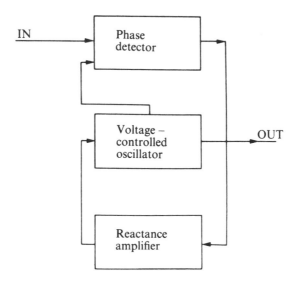

Figure 7.21   Block diagram of the phase-locked loop

an oscillator whose frequency can be controlled by the value of current (or voltage, according to design) flowing into (or being applied to) a reactance amplifier. The oscillator output is compared with that of an input signal. If this input signal is within a certain range (the so-called *capture band*) of the oscillator frequency, control of the oscillator is taken over by the d.c. voltage generated by the comparator (or phase detector).

The reactance amplifier is a transistor stage whose action is made to resemble that of a reactance (current 90° out of phase with voltage) by positive feedback through a reactor. The circuit action as a whole will be familiar to the TV engineer as being that of the flywheel sync circuit. The versatility of the circuit is increased by the fact that a d.c. correction signal can be obtained from it, in addition to the corrected oscillator signal.

The PLL can be used simply as a precision oscillator whose frequency is set by the values of a single resistor and capacitor, as in Figure 7.22. In this application, the oscillator frequency is stable to within 60 parts per million (which means 60Hz in 1MHz), provided that a stabilized supply voltage is used. The value of C_2, is made ten times that of C_1, and the output frequency is given by the equation: −

$f_o = \dfrac{1}{R_1 C_1}$ where R_1 is in ohms and C_1 in farads.

Another mode of employment is as a very versatile detector and filter. The internal oscillator frequency can be set to practically any frequency, within wide

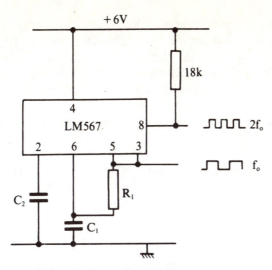

Figure 7.22   The PLL as a precision oscillator

limits, and the bandwidth over which the internal oscillator will lock to the frequency of the incoming signal can be set independently between limits of 0% and 14% of the frequency of the oscillator. With the bandwidth control set to minimum, the oscillator will lock very precisely to a frequency selected by the timing components.

For example, if the input to the PLL is a crystal-controlled square wave, the timing components (R_3 and C_2 in Figure 7.23) can be set so that the output

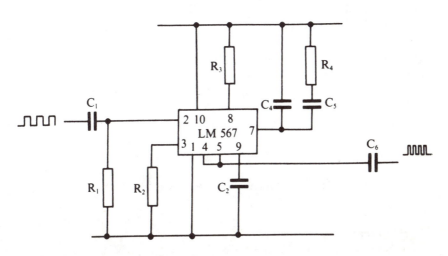

Figure 7.23   The PLL as a detector and filter

197

frequency is at one of the harmonics of the frequency of the crystal (provided, of course, that this does not exceed the frequency range of the PLL). Every harmonic selected in this way will be perfectly locked in phase to the crystal frequency.

Such an arrangement is used in frequency-synthesiser circuits, since it can combine the versatility of the variable-frequency LC oscillator with the stability of crystal control.

In addition, the components C_4-C_5-R_4 in the Figure 7.23 circuit form a filter circuit. Component values for this function can be calculated from the manufacturers' data sheet.

A major advantage of PLL circuits is that they make it possible to regenerate a signal which is practically lost in noise. Provided the input signal amplitude is enough to drive the PLL (only about 20mV is needed in the NE567), any input frequency which is within the capture range of the frequency set by the timing capacitor and resistor will be locked in, and the output will be a waveform of that frequency which is free from both noise and interference.

A frequency-modulated signal can be demodulated, even if the signal is varying considerably in amplitude, because a change in the frequency of the input signal causes a change in the steady voltage of the phase detector signal. Since the voltage thus fluctuates exactly according to the frequency variations, it becomes the demodulated signal itself.

Exercise 7.6

Examine the data sheet for a PLL. If a suitable IC is available, construct a test circuit as follows.

Select values for the timing components so as to give an oscillator frequency of 10kHz, and for the bandwidth capacitor to give a bandwidth of 1kHz. Follow the procedure laid down in the data sheet to select a value for the output filter capacitor.

Using an oscilloscope to monitor the output, feed to the input a signal of about 50mV (r.m.s.) from a signal generator. Vary the frequency of the signal generator from 5kHz to 15kHz, and observe the frequency and amplitude of the output signal.

The Schmitt trigger

The Schmitt trigger is a circuit having three principal functions:

1 To sharpen up the shape of a pulse which has become integrated after transmission through a long cable.
2 To turn sine-waves into square waves.
3 To detect a pre-set level of voltage.

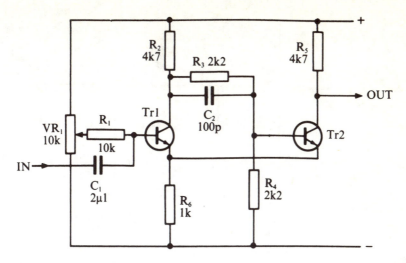

Figure 7.24   The Schmitt trigger circuit

A simple such circuit is shown in Figure 7.24. VR_1 can be adjusted so that Tr1 is either cut off or bottomed. Assume that the control has been set so that Tr1 is cut off. Tr2 conducts because of the current flowing through R_2 and R_3 into its base. With Tr2 conducting, a voltage will exist across R_6 – approximately 2.9V in the circuit shown. Tr1 cannot conduct until its base voltage is at least 0.5V higher than this – i.e., about 3.4V. With the emitters of the two transistors coupled by R_6 and the collector of Tr1 connected to the base of Tr2, a positive feedback loop exists for d.c. signals which will come into action whenever both transistors conduct.

A change in the setting of VR_1 to more than 3.4V will cause the circuit to flip over, with Tr1 now conducting and Tr2 cut off. The potential divider action of $R_3 - R_4$, however, will not allow the circuit to switch back at the same value of voltage from VR_1 as was needed to switch over. The difference between the switchover and the switch-back voltages is called *voltage hysteresis*. The amount of hysteresis is determined by the values of R_3 and R_4 in the circuit shown.

The level of output voltage in the Schmitt trigger is either high or low, according to whether the circuit is switched over or not. VR_1 sets the threshold voltage at which triggering will occur in one direction, and from any input signal large enough to trigger the circuit a square-sided output signal will be generated.

The capacitor C_2 is a 'speed-up' capacitor which has the effect of sharpening the sides of the square pulses by compensating for the integrating effects of stray capacitance.

A sketch of the action of a Schmitt trigger is shown in Figure 7.25. The width of the rectangle indicates the amount of input voltage hysteresis, and the levels of the stable maximum and minimum output voltages are shown as its top and bottom sides.

199

Servicing Electronic Systems Volume 2, Part 1

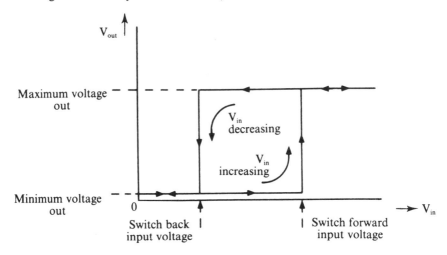

Figure 7.25   Schmitt trigger action

Exercise 7.7

Construct the Schmitt trigger circuit shown in Figure 7.26. Both Tr1 and Tr2 are

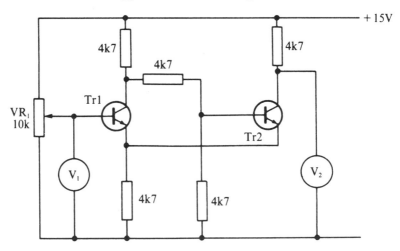

Figure 7.26   Circuit for Exercise 7.6

general purpose NPN types. Set the potentiometer VR_1 so that the base voltage of Tr1 will be zero when the circuit is switched on.

Switch on, and note the readings on the two voltmeters, both of which should be set to the 0–15V range. Then adjust VR_1 so that the reading of voltmeter V_1 increases to 0.5V. Note this reading, and the simultaneous reading of V_2. Continue

200

to take and note both readings as VR_1 is so adjusted that the voltage readings of V_1 increase in steps of 0.5V.

Determine the voltages at which the circuit flips over, in both directions.

Now remove the voltmeters, connect a 10µF input capacitor to the base of Tr1 and connect an oscilloscope to the collector of Tr2. Adjust VR_1 to give a voltage at the base of Tr1 which is midway between the flip-over and the flip-back voltages. Connect a signal generator to the input capacitor and set it to a frequency of 400Hz, sine-wave.

Examine the output on the oscilloscope as the amplitude of the signal from the generator is gradually raised to the flip-over point of Tr1.

Measure and tabulate the d.c. voltage levels of the emitter, base and collector of each transistor with the circuit in each of its two possible conditions.

The Miller integrator

The Miller integrator is a circuit based on the application of negative feedback through a capacitor.

In the circuit of Figure 7.27, C_1 (the so-called *Miller capacitor*) is connected

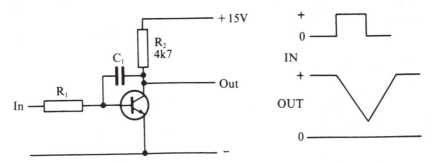

Figure 7.27   The basic Miller integrator circuit, with waveforms

between the collector and the base of a transistor. A rise of positive voltage at the input large enough to start the transistor conducting would, if C_1 were not there, cause a large negative change in voltage at the collector. Negative feedback through C_1, however, prevents the voltage at the base of Tr1 from rising faster than the rate at which C_1 can charge through R_1.

This charging rate does not follow the usual exponential shape (while the transistor is conducting) because of the feedback through C_1 and a good linear sweep waveform is obtained.

The op-amp version of the Miller integrator has already been explained under the heading of The op-amp integrator (p. 108).

The principal application of the Miller integrator is in timebase circuits for oscilloscopes. If the gain of the transistor is high, the sweep waveform can be made

201

very linear indeed.

A complete Miller timebase circuit for an oscilloscope is made considerably more elaborate than the basic circuit shown by the addition of other circuits whose purposes are (a) to discharge the Miller capacitor rapidly (so obtaining rapid sweep flyback), and (b) to switch the input voltage at the times required to generate the timebase waveform.

Exercise 7.8

Construct the simple Miller integrator of Figure 7.27 giving R_1 a value of 1k and C_1 a value of 1μF. Use a signal generator set to inject into the circuit square waves at 500Hz, 1.5V peak-to-peak, and observe the voltage waveform at the collector with the aid of an oscilloscope.

Making no other changes, try the effects of the following set of values:

(a) $R_1 = 10k$; $C_1 = 1μF$
(b) $R_1 = 1k$; $C_1 = 0.1μF$.

Note that arrangement (a) gives C_1 a time constant ten times what it was before, and that arrangement (b) gives it a time constant which is one tenth of the previous value. Note the effects of these changes, and sketch the waveforms.

Unijunction oscillators

The unijunction has already been briefly mentioned. Figure 7.28 shows an oscillator

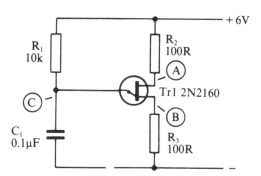

Figure 7.28   The unijunction oscillator

circuit in which one is used.

At switch-on, the emitter of the unijunction is held at zero voltage by C_1; but as C_1 charges through R_1, emitter voltage rises until the device triggers. It will be recalled that triggering takes place when the emitter voltage equals $k \times V_s$ where k is the quantity known as the *intrinsic standoff ratio*, which is a constant for a given unijunction, and V_s the supply voltage.

When the unijunction triggers, C_1 discharges very rapidly through the unijunction, so that a large pulse of current flows through R_3 and also (because the junction is conducting so well) through R_2. The waveforms produced are a charge-discharge sawtooth at the emitter, with short pulses in antiphase at the two base electrodes.

The time period between pulses depends on the values of R_1 and C_1 and of k, but is independent of the supply voltage.

Unijunctions are much used as pulse and sawtooth generators, particularly to trigger the firing of thyristors.

Exercise 7.9

Construct the unijunction shown in Figure 7.28. Switch on, and with an oscilloscope measure the amplitude and the time period between pulses of the waveforms at Points A, B, and C. Sketch these waveforms. Measure the d.c. voltage levels at the points A, B and C.

Now determine the effects on the waveforms of the following changes:

(a) Replace the 0.1µF capacitor by one having a value of only 0.01µF.
(b) Put back the original 0.1µF capacitor, and replace the 10k resistor by one having a value of 100k.

The 555 timer

The 555 timer is an IC which is widely used as an oscillator or as a generator of time delays. The timing depends on the addition of external components (a resistor and a capacitor), and one very considerable advantage of using the IC is that the time delays or waveforms that it produces are well stabilized against changes in the d.c. supply voltage. As with other ICs, we shall ignore the internal circuitry and concentrate on what the chip does.

The chip pin layout for the usual 8-pin DIL form is shown in Figure 7.29, using pins 1 and 8 for earth and supply positive respectively. The output is from pin 3, and the other pins are used to determine the action of the chip. Of these, pins 2, 6 and 7 are particularly important. Pin 7 provides a discharge current for a capacitor that is used for timing, and pin 6 is a switch input that will switch over the output of the circuit as its voltage level changes. For most uses of the chip, these pins are connected to a CR circuit whose charge and discharge determines the time delay or the wavetime of the output.

Figure 7.30 shows the timer used to operate a relay for a time which can be selected by the setting of switch Sw2 and a 470k potentiometer. When the switch Sw1 is momentarily pushed, pin 2 is earthed, causing the output from pin 3 to go

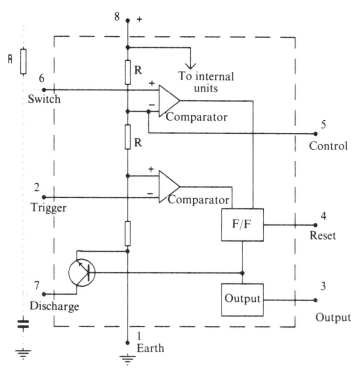

Figure 7.29 **The internal arrangement and pin connections for the 555 timer IC**

high and operate the relay. At the same time, pin 7, which has been connected internally to earth, is isolated, so that the timing capacitor selected by Sw2 can start to charge through the resistors (100k in series with the variable 470k). When this voltage reaches a value of about two-thirds of supply voltage, the input to pin 7 will cause the circuit to flip over again, releasing the relay and discharging the timing capacitor. The action is then complete until the switch Sw1 is pressed again. The diodes D_1 and D_2 are needed to protect the IC against the large pulse of voltage which occurs when the relay is switched off. D_1 can be any general-purpose silicon diode, and D_2 should be a fast-acting diode with a small forward voltage drop, such as a gold-bonded germanium diode or a Schottky silicon diode.

An astable pulse generator circuit is illustrated in Figure 7.31. Pin 6 is used to discharge the capacitor C and is connected to the triggering pin 2. This ensures that the circuit will trigger over to discharge the capacitor when the voltage level reaches two-thirds of supply voltage, and will trigger back to allow the capacitor to charge when the voltage level reaches one-third of supply voltage. The frequency of the output can be adjusted by using the 100k variable and, if needed, by using switched values of capacitance.

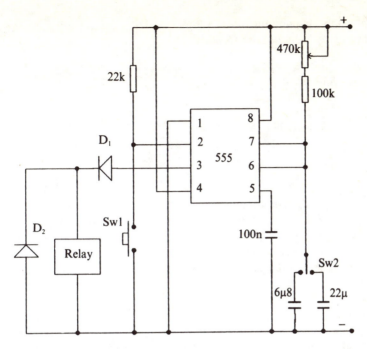

Figure 7.30 **A circuit that uses the 555 timer to generate a relay switching pulse of variable time**

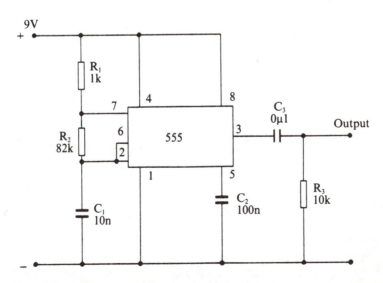

Figure 7.31 **An astable circuit using the 555 timer to generate the square wave**

For the astable circuit illustrated in Figure 7.31 the waveform timings are given approximately by the following formulae:

Charge time $0.7 (R_1 + R_2)C_1$
Discharge time $0.7 R_2C_1$
Periodic time $0.7 (R_1 + 2R_2)C_1$
Frequency $\dfrac{1.44}{(R_1 + 2R_2)C_1}$

Exercise 7.10

Construct an astable oscillator using the 555 timer with the component values shown in Figure 7.31. Use the oscilloscope to view the output waveform and measure its frequency. Sketch this waveform, indicating peak-to-peak voltage and periodic time. View and sketch also the waveform at pin 6, showing the voltage levels and the time for each part of the waveform. Measure the current taken by the circuit.

Exercise 7.11

ICs, either op-amps or digital types, can be used as the basis of oscillators, and the following exercise uses a digital gate circuit (see Chapter 10 for details) as the basis for a crystal-controlled oscillator. The circuit is shown in Figure 7.32a, showing the connections to the chip. Figure 7.32b shows the actual oscillating circuit which uses two of the NAND gates (see Chapter 9) in the chip. NOTE that the supply voltage MUST BE 5V – a stabilized supply should be used and its voltage checked before connecting the circuit.

With the crystal removed so that there is no oscillation, measure the d.c. voltage levels at pins 1 and 2, pin 3, pins 4 and 5 and pin 6.

Now replace the crystal and use an oscilloscope connected to the output of pin 6 to measure the peak-to-peak amplitude and the periodic time of the output.

Measure also the d.c. voltage levels on the pins, keeping the oscilloscope connected so that you can monitor whether the oscillator continues to oscillate when the meter is connected.

Multiple-choice test questions

1 Which of the following is not a requirement for an oscillator circuit?
   (a) positive feedback
   (b) negative feedback
   (c) voltage gain
   (d) time constants.

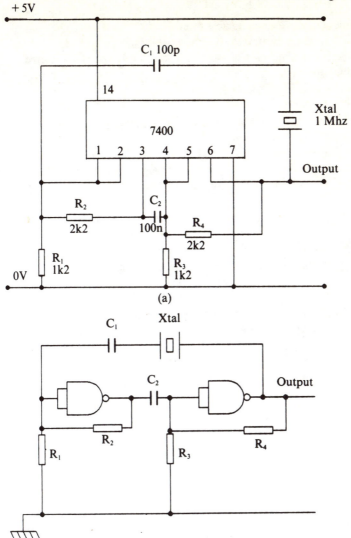

Figure 7.32 **(a) IC pin diagram, (b) actual circuit for Exercise 7.10**

2 The Colpitts and Hartley circuits are examples of:
   (a) astable multivibrators
   (b) class A bias
   (c) sine-wave oscillators
   (d) bistables.

3 A unique feature of an astable multivibrator is that:
   (a) it can be synchronized to a range of input frequencies
   (b) its frequency is perfectly stable when supply voltage is altered
   (c) its output is a sine wave
   (d) its output is a perfect square wave.

4 The important feature of a phase-locked loop (PLL) is that:
   (a) it provides a stable output frequency
   (b) its output frequency will be locked to an input frequency
   (c) its output frequency can be determined by a crystal
   (d) its output frequency is not affected by changes in supply voltage.

5 A Schmitt trigger circuit's main action is to:
   (a) convert a square wave into a sine wave
   (b) convert a square wave into a triangular wave
   (c) operate as an oscillator generating a square wave
   (d) create a steep-sided wave from a slowly-varying input.

6 The Miller integrator, given a square wave input, will generate:
   (a) a sinewave
   (b) another squarewave
   (c) a sawtooth or triangular wave
   (d) a set of pulses.

8 Transformers and transducers

Summary

Ideal transformer, transformer equations. Transformer types. Losses. Constructional details. Screening. Faults. Transducers. Photocells. LEDs. Opto-couplers. Thermistors. Thermocouples. Microphones and Loudspeakers.

Transformers

A transformer consists of two or more inductors so wound that their magnetic fields interact. (Note, however, that that definition does not hold good of the auto-transformer – for which see below.)

Methods of causing the inductors to interact, or *couple*, in this way are to wind the coils very close to one another, and/or to put magnetic cores into the coils themselves.

The windings of a transformer are called *primary* and *secondary* respectively. Signals, which may be either a.c. or unidirectional, are applied to the primary winding(s) and induce output signals, which are always a.c. – never unidirectional – across the terminals of the secondary winding(s). A steady d.c. current flowing through the primary windings induces no output signal in the secondary windings at all.

The signal current flowing through the primary winding gives rise to a fluctuating magnetic field (ideally, a graph of magnetic field plotted against time would have exactly the same wave-shape as would a graph of signal current in the primary winding, also plotted against time). This fluctuating magnetic field induces a signal voltage in the secondary winding.

When the current is taken from the secondary winding by connecting a circuit to it, increased primary current must flow to provide the power which is being dissipated. If no current is taken from the secondary winding, current flow in the primary winding should be very small.

Some transformer constructions are illustrated in Figure 8.1. Type (a) would

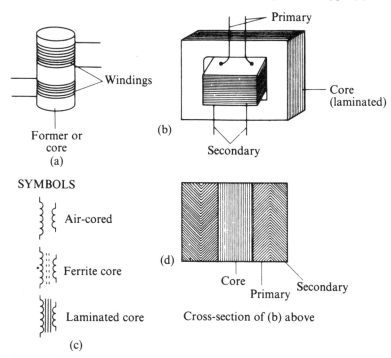

Figure 8.1   Transformer types and symbols

be used at radio frequencies; type (b) at lower frequencies – with a cross-section through the construction shown at (d). Figure 8.1(c) gives the symbols used to indicate various types of transformer core.

An ideal transformer

The ideal transformer would be one which suffers no loss of power when in use, so that no primary current at all would flow until secondary current was being drawn off. Large transformers in fact approach quite close to this ideal which is used as the basis of all transformer calculations. In an ideal transformer:

$$\frac{V'_s}{V'_p} = \frac{n_s}{n_p}$$

where V'_s = a.c. voltage, secondary; V'_p = a.c. voltage, primary; n_s = number of turns of wire in the secondary winding; and n_p = number of turns of wire in primary winding.

Example. A transformer has 7500 turns in its primary winding and is connected to a 250V 50Hz supply. What a.c. voltage will be developed across its secondary winding if the latter has 500 turns?

Solution. Substitute the data in the formula:

$$\frac{V'_s}{V'_p} = \frac{n_s}{n_p}$$

$$\text{Then } \frac{V'_s}{250} = \frac{500}{7500} = \frac{1}{15}$$

$$\text{So that } V'_s = \frac{250}{15} = 16.67V, \text{ or say about } 17V.$$

In practice, because no transformer is perfect, the output voltage would be somewhat less than 16V.

In the ideal transformer, the power input to the primary winding must be equal to the power taken from the secondary winding, so that:

$$V'_p . I'_p = V'_s . I'_s \text{ or, re-arranging, } \frac{V'_s}{V'_p} = \frac{I'_p}{I'_s}$$

$$\text{and since } \frac{V'_s}{V'_p} = \frac{n_s}{n_p}, \text{ it follows that } \frac{I'_p}{I'_s} = \frac{n_s}{n_p}$$

The latter equation can also be expressed as $I'_p . n_p = I'_s . n_s$, an often convenient form which relates the signal current flows in the perfect transformer to the number of turns in each of the two windings.

Transformers are used in electrical circuits for the following purposes:

1   Voltage transformation – converting large signal voltages into low voltages, or low signal voltages into large ones, with practically no loss of power.
2   Current transformation – converting low-current signals into high-current signals, or vice versa, with practically no loss of power.
3   Impedance transformation – enabling signals from a high-impedance source to be coupled to a low impedance, or vice versa, with practically no loss of power through mismatch.

Note carefully, however, that the transformer is a passive component which gives no power gain. If a transformer has a voltage step-up of ten times, it will also have a current *step-down* of ten times (assuming no losses en route).

Example. The secondary winding of a transformer supplies 500V at 1A. What current is taken by the 250V primary?

Solution. Since $V'_s.I'_s = V'_p.I'_p$
then $250 \times I'_p = 500 \times 1$,
$\therefore I'_p = 2A$

The use of a transformer in impedance matches has been discussed in Chapter 5; the matching condition is:

$$\frac{R_s}{R_p} = \left(\frac{n_s}{n_p}\right)^2$$

This relationship can be usefully re-expressed in the form:

$$R_p = \left(\frac{n_p}{n_s}\right)^2 \times R_s$$

so that the equivalent input resistance to signals entering a perfect transformer is

$$R_L \left(\frac{n_p}{n_s}\right)^2$$

where R_L is the load resistance connected to the secondary winding. The equivalent circuit for a perfect transformer is therefore that shown in Figure 8.2.

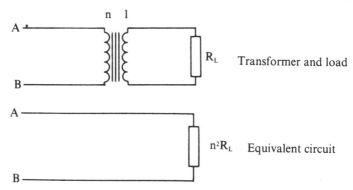

Figure 8.2   Equivalent circuit of a perfect transformer

Transformer losses

The types of power loss which a transformer can suffer are the following:

1 I^2R losses caused by the resistance of the windings
2 Eddy-current and stray inductance losses caused by unwanted magnetic interactions

3   Hysteresis loss arising from the core material, if a core is used.

Taking these in turn, I²R (or *joule*) losses are those which are always incurred in any circuit when a current, steady or a.c., flows through a resistance. These losses can be reduced in a transformer by making the resistance of each winding as low as possible consistent with the correct number of turns and the size of the transformer.

Joule losses are generally insignificant in transformers used at radio frequencies; but they will cause overheating of mains transformers, particularly if more than the rated current is drawn or if ventilation is inadequate.

Stray inductance and *eddy current losses* are often more serious. An ideal transformer would be constructed so that all the magnetic field of the primary circuit coupled perfectly into the secondary winding. Only toroidal (ring-shaped) transformers come close to this ideal (and even those in the smaller sizes only). In practice, there arises from the primary winding a strong alternating field which is detectable at some distance from the transformer, causing a loss of energy by what is termed stray inductance.

In addition, the alternating field of the primary can cause stray voltages to be induced in any conducting material used in the core or casing of the transformer, so that unwanted currents, called eddy currents flow. Since additional primary current must flow to sustain these eddy currents, they cause a loss of power which can be significant.

The problem of eddy currents in the core is tackled in two ways:

1   The core is constructed of thin laminations clamped together, with an insulating film coating on each to lessen or eliminate conductivity
2   The core is constructed from a material possessing high resistivity, such as ferrite.

The third type of loss, called *hysteresis loss*, occurs only when a magnetic core is used. It represents the amount of energy which is lost when a material is magnetized and de-magnetized. This type of loss can be minimized only by careful choice of the core material for any particular transformer.

Hysteresis losses will, however, increase greatly if the magnetic properties of the core material change, or if the material becomes magnetically saturated. The following precautions should therefore be taken in connection with transformers:

1   Do not dismantle transformer cores unnecessarily, nor loosen their clamping screws
2   Never bring strong magnets near to a transformer core
3   Never pass d.c. through a transformer winding unless the rated value of the d.c. is known and is checked to be correct.

Transformer construction

The way in which transformers are constructed depends greatly on the frequency range for which they are intended. The following conditions are samples only of the considerations which may apply.

Mains supply   Mains frequency is low (50 to 60Hz) and fixed, and a substantial core of silicon-iron is required. The core must be laminated, and hysteresis loss can be reduced to negligible proportions by careful choice of a core material. Where an external magnetic field is especially undesirable (as in audio amplifiers and cathode ray oscilloscopes), a toroidal core can with advantage be used.

Audio-frequency range   The core material must be chosen from materials causing only low hysteresis loss because of the higher frequencies which will be encountered, and the windings must be arranged so that stray capacitance between turns is minimized. In general, any flow of d.c. is undesirable.

Lower-range radio frequency   In this range, the losses from laminated cores are unacceptably high, so that ferrite cores must be used. Because of the high frequencies involved, a small number of turns is sufficient for each winding. Stray fields are difficult to control, so that *screening* (see below) is often needed.

Higher-range radio frequency   Only air cores can be used in this range, and 'coils' may actually comprise less than one full turn of wire. They may even consist of short lengths of parallel wire. Unwanted coupling becomes a major problem, so that the physical layout of components near the transformer assumes great importance.

Two variations on transformer construction are *tapping*, and *bifilar winding*.

A tapping is a connection made to a selected part of the wire of a winding so that different numbers of turns may be selected for different jobs. Mains transformers, for example, will have several tappings on the primary winding so as to be able to handle different values of mains voltage. Multiple-tapping secondary windings (Figure 8.3(b)) can also be used to provide several different ratios from one transformer, so that different secondary voltages or matching ratios can be obtained.

A centre-tapping secondary winding (Figure 8.3(a)) is a good way of obtaining phase-inverted signals for audio amplification or for rectification. This principle can be extended to the auto-transformer, which is a single-tapped winding equivalent to the use of a double-wound transformer with one end of

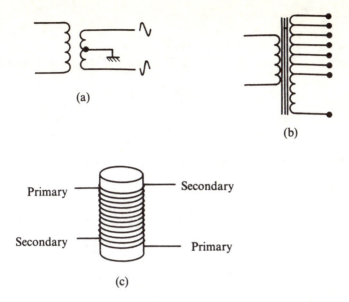

Figure 8.3   Types of transformer windings

the primary connected to one end of the secondary. The ratio of input/output voltages and currents still follows the normal transformer relationships. An auto-transformer with a variable tapping position (such as the Variac) is used for providing variable voltage a.c. supplies. Note, however, that such transformers provide no isolation between their primary and secondary windings.

Bifilar winding (Figure 8.3(c)) is a method of providing very close coupling between primary and secondary windings, particularly useful in audio transformers. In this method of construction, the primary and the secondary turns are wound together, rather than in separate layers.

Shielding or screening

Components such as transistors and inductors may need to be shielded from the fields which are radiated by transformers. Electrostatic screening is comparatively easy, for any earthed metal will screen a component from the electrostatic field of a transformer (though with high frequencies a metal box which is almost watertight may have to be used).

Electromagnetic screening, on the other hand, calls for the use of high-permeability alloys such as *mu-metal*, or *super-Permalloy*. Surrounding a component with such a material ensures that no magnetic fields from outside the box can penetrate into it.

Exercise 8.1

Using a transformer of known turns ratio, preferably a type using a tapped secondary winding, connect the circuit shown in Figure 8.4. Measure the a.c.

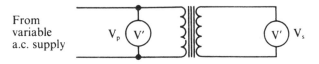

Figure 8.4   Circuit for Exercise 8.1

input and output voltages for each set of taps, and find the values of V'_s/V'_p. Compare these values with the known values of the turns ratio.

An effective component to use in this experiment is a toroidal core with a 240V primary winding, obtainable from most educational suppliers.

Transformer faults

The following are common transformer faults, with hints on how to detect and remedy them:

- Open-circuit windings, which can be detected by ohmmeter tests. A winding may also acquire high resistance, typically 100k instead of 100 ohms.
- Short-circuit turns, which are difficult to detect because the change of resistance is very small. S/c turns will cause an abnormally large primary current to flow when the secondary is disconnected, so that mains transformers overheat and transformers operating at high frequencies fail completely. This is a fault which particularly affects TV line output transformers. The most certain test and cure is replacement by a component known to be good.
- Loose, damaged or missing cores. Loose cores will cause mains transformers to buzz and overheat. Cracked or absent ferrite cores in radio-frequency transformers will cause mis-tuning of the stage in which the fault occurs.

Summary

A transformer transfers signals (but not their d.c. level) from a primary winding to a secondary winding. The signal voltage can be stepped up or down, depending on the relative number of turns used in the two windings. Impedance matching can also be achieved by means of a transformer.

The type of construction of a transformer, and its core material if used, must be carefully chosen to suit the range of frequencies which the transformer is being designed to handle.

Transducers

A transducer is a device that converts energy from one form to another. The transducers that are of most interest as far as electronics is concerned are those in which one of the energy forms is electrical energy. Of the many possible transducers, the most common are the transducers for light, heat and sound.

Photo-electric transducers

As the title suggests, these devices convert light energy into electrical signals. They are used extensively in position sensing and counting devices where the interruption of a light beam is converted into an electrical signal. As most of these devices have a wavelength response greater than the human eye, their applications extend into the infra-red region. Figure 8.5 shows the typical human-eye response in relationship to the infra-red wavelengths.

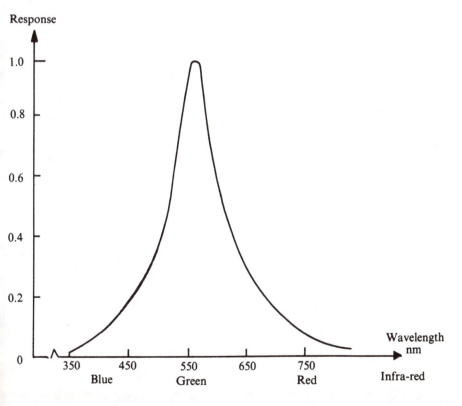

Figure 8.5   Response of human eye to visible light

Photo-conductive cells

The resistance of certain materials decreases when subjected to increased illumination. The common materials used for light-dependent resistors (LDRs), as these devices are called, include: selenium, cadmium sulphide, cadmium selenide and lead sulphide. Each material responds to different wavelengths of light. Figure 8.6(a), shows the general appearance of a photo-conductive cell and its

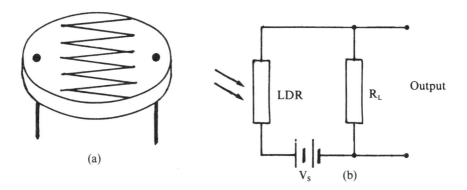

Figure 8.6 **(a) The photo-conductive cell and (b) basic circuit**

construction. Gold-conducting sections are deposited on a glass plate with a long, meandering gap to isolate the two sections. A thin layer of suitable photo-conductive material is then deposited to bridge the insulating region. This construction is necessary to reduce the cell's resistance to a usable value. Typical resistance values for various cells range from about 500kΩ to 10MΩ in darkness and from about 1kΩ to 100kΩ in bright light. The change in resistance is non-linear and there is a significant time lag of up to 0.1 sec. in response to a pulse of light. Figure 8.6(b) shows a basic circuit using a LDR. The current flowing in the load resistance R_L due to the supply voltage V_s produces the output voltage. Increasing the illumination of the LDR lowers its resistance so that the current, and hence the output voltage, increases. The maximum permitted voltage for these cells may be as high as 300v and the maximum power dissipation is in the order of 300mW. The dark-to-light resistance ratio ranges from about 50:1 to 250:1. The cells are particularly sensitive to red light and the infra-red wavelengths, so that they are often used as flame detectors in boiler and furnace control systems.

Photo-diodes

Photo-diodes are formed with one very thin region and equipped with a lens so that light energy can be directed into the depletion region. When this happens 'hole-electron' pairs are generated to increase the diode's conductivity. Such diodes have a peak response in the infra-red region, but the response to visible light is still very useful. The diodes are operated either reverse-biased, or only very slightly forward-biased (to increase sensitivity) so that no current flows. Typical 'dark current' is as low as 2nA and rising to about 100µA in bright light for germanium types. Modern PIN silicon versions may have a peak power dissipation greater than 100mW. The increase in current due to the light is practically linear. The response time to a pulse of light is very short so that they find applications in high-speed switching circuits. Generally, the small output is a disadvantage and amplification is needed. The basic circuit configuration is shown in Figure 8.7(b) whilst the more practical application of a light meter is

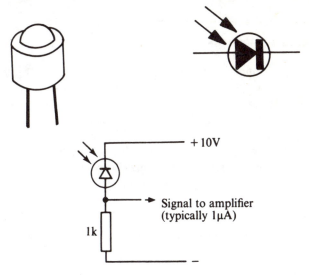

+10V

Signal to amplifier
(typically 1µA)

1k

Figure 8.7   (a) Photo-diode and circuit symbol and (b) a silicon photo-diode (input circuit)

shown in Figure 8.8. With both diodes shielded from light, the conduction of the two transistors is balanced by resistor R so that there is no current flowing through the meter. In operation D_2 is maintained in darkness and the base voltage of Tr1 and hence its collector current depends on the level of light falling on D_1. An increase in the illumination of D_1 causes Tr1 collector current to decrease, its collector voltage rises and so the meter deflects from its zero position.

Photo-transistors

A photo-transistor has a similar construction to a silicon planar transistor except that it is equipped with a lens so that light can be made to shine directly

219

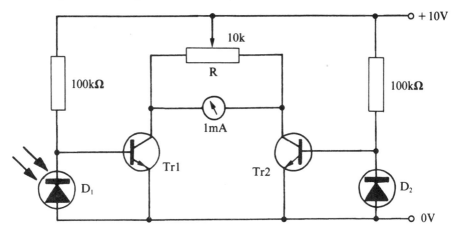

Figure 8.8   Circuit of light meter

into the base-emitter junction. Their response time to a pulse of light is in the order of 1-2μsec and are therefore mostly used in medium-speed light detector circuits. The current ranges from about 25 to 50μA in darkness to about 5 to 10mA in bright light. The photo-transistor is connected so that the base is either open-circuit, or reversed-biased, or only very slightly forward-biased. In some devices, the base connection is omitted altogether so that only the collector and emitter connections can be used. The load may be placed in either the collector or emitter leads as shown in Figure 8.9. Light shining into the base-emitter

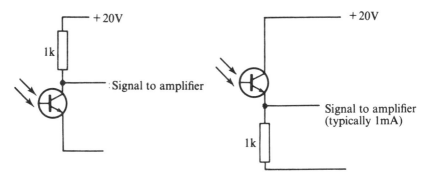

Figure 8.9   A photo-transistor as a light-level detector

junction modulates the base current which is in turn magnified by the current gain. Such a device can be used to drive a load directly as shown in Figure 8.10.

Light emitting diodes (LEDs)

The operation of the photo-diode was shown to be dependent upon the application of energy to generate hole–electron pairs. Conversely, when holes and

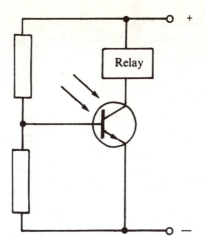

Figure 8.10   A photo-transistor as a relay driver

electrons re-combine, energy is released. In germanium and silicon, this energy is released as heat into the crystal structure. However, in materials such as gallium arsenide and gallium phosphide this energy is released as light, different semiconductor compounds releasing light of different wavelengths. The basic structure of the diode is shown in Figure 8.11(a). The plastic moulding not only holds the component parts together but also acts as a light-pipe so that most of the light generated is radiated from the domed region. A characteristic typical of

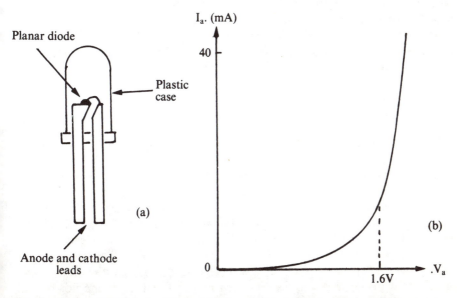

Figure 8.11   (a) Construction of LED and (b) characteristic of LED

221

a light-emitting diode is shown in Figure 8.11(b). When forward-biased beyond about 1.6 volts, the current rises rapidly and light is released. The current should be limited to a value less than about 40mA by the use of a series resistor.

Opto-couplers

These consist of an LED and photo-transistor pair as shown in Figure 8.12. The

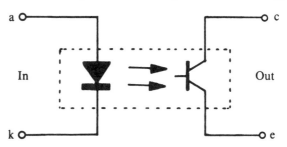

Figure 8.12   Opto-coupler

input signal modulates the diode current and hence the intensity of its light output. This variation in light produces a variation in collector current to provide an output signal. Since the light beam has no electrical impedance, there is no matching problem between input and output. The electrical isolation is very high and an opto-coupler can withstand test voltages as high as 4kV between input and output terminals. Alternatively the LED can be used to launch energy into a glass optical fibre cable to transmit the signal over very much greater distances. The light path within the opto-coupler can be interrupted by a shutter. This generates a pulsating signal that might be used to drive a counter circuit capable of working at high speed.

The photo-transistor may be replaced by a compound transistor (Darlington amplifier) to provide higher gain or by a photo-thyristor to provide higher output current. However, both arrangements result in a lower switching speed.

Photovoltaic devices

A layer of selenium is deposited on an iron or aluminium backing plate which forms the positive pole. A transparent layer of gold is evaporated on to this to form the negative pole. A metallic contact ring completes the circuit. The general construction and circuit symbol is shown in Figure 8.13. Light shines through the gold film into the layer of selenium. This releases electrons that form an electric field within the selenium, making the gold layer the negative pole of the cell. The whole cell is enclosed in a plastic housing with a transparent window for protection. The maximum current in bright light depends on the particular cell, but short-circuit currents in excess of 1mA can be obtained. Such

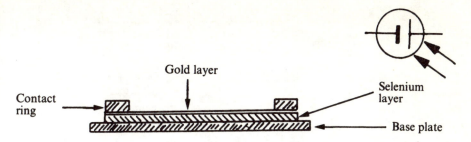

Figure 8.13   Section through photo-voltaic cell and its circuit symbol

cells are suitable to drive a portable photographic light meter.

Solar cells

These are the semiconductor version of the selenium cell. They are formed from heavily doped PN junctions. A cell of about 4 sq.cm. is capable of providing 0.6V on open circuit, with a short-circuit current of up to 100mA in bright light.

Thermistors

These devices may be in rod, bead, washer or disc form. They are made from carefully controlled mixtures of certain metallic oxides. These are sintered at very high temperatures to produce a ceramic finish. Thermistors have a large *temperature coefficient of resistance*, given by the expression:

$$\frac{\text{Change of resistance}}{\text{original resistance} \times \text{change of temperature}}$$

In normal resistors, this value is relatively small, typically in the order of 3×10^{-3}. For thermistors however, the temperature coefficient can be of the order of $(15 \text{ to } 50) \times 10^{-3}$. Negative temperature coefficient (NTC) thermistors have resistances that fall with a rise in temperature and are commonly made from mixtures of the oxides of manganese, cobalt, copper and nickel. Positive temperature coefficient (PTC) components whose resistance increases with a rise in temperature can be made from barium titanate with carefully controlled amounts of lead or strontium.

Thermistors are used extensively for temperature measurement and control up to about 400°C. A typical circuit is shown in Figure 8.14 where the bridge circuit is balanced at some low temperature so that the output from the differential amplifier is zero. Th_2 is maintained at this low temperature, whilst Th_1 is exposed to the temperature to be measured. A rise in the temperature of Th_1 causes its resistance to rise and lower the voltage at one of the amplifier inputs. The amplifier output now changes in proportion to the rise in temperature. The

223

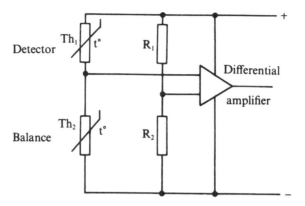

Figure 8.14   A thermistor bridge

addition of a thermistor to the bias network of an amplifier can be used to stabilize it against the effects of temperature change. Another application is to provide temperature compensation for the change in the winding resistance of alternators and other generators which affects their performance when the operating temperature rises.

Thermo-couples

When two dissimilar metals are in contact with each other a *contact potential* is developed between them. This is known as *Seebeck* or *thermo-electric* effect. The voltage, which rises with temperature, is almost linear over several hundred degrees. If two junctions are formed as shown in Figure 8.15(a), a current will

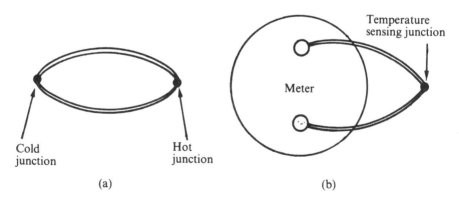

Figure 8.15   (a) and (b) Principle of thermo-couple

flow around the circuit provided that each junction is at a different temperature. The circuit can be modified as shown in Figure 8.15(b) to include a meter which now becomes one of the junctions, the other becoming the temperature sensor. By reversing the meter connections, temperatures below ambient can be measured. The two metals are chosen to maximize the contact potential for a particular temperature range. The metals used include:

- iron and copper/nickel alloy,
- copper and copper/nickel alloy,
- nickel/chromium alloy and copper/nickel alloy,
- nickel/chromium alloy and nickel/aluminium/manganese alloy.

These allow for a range of instruments to measure temperatures from $-85°C$ up to 2000°C. For most temperature measurements, the hot junction is formed into a probe as shown in Figure 8.16. The metal sheath is used to protect the junction

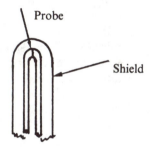

Figure 8.16   A typical high-temperature probe

from environmental hazards and is electrically isolated from it with magnesium oxide. This material has good thermal conductivity with high electrical resistance (up to $1000M\Omega$).

Microphones and loudspeakers

Energy conversion between sound and electrical waves is carried out by the familiar transducers, the microphone and the loudspeaker. Sound is a wave motion in the air (or in other, more dense, materials), and is the result of vibration. The vibration of a material which is in contact with the air will create a sound wave, and when a sound wave strikes any material it will set it into sympathetic vibration. This action of the sound wave is the way in which sound is detected by our own eardrums and by microphones.

Microphone types are classified by the type of transducer they use, and by several other characteristics, of which impedance is particularly important. A microphone with high impedance usually has a fairly high electrical output, but the high impedance makes it very susceptible to hum pickup. A low impedance microphone will have a very low electrical output, but hum pickup is almost

225

negligible.

Another factor of importance is whether the microphone is directional or omnidirectional. If the transducer senses the pressure of the sound wave, the microphone will be omnidirectional, picking up sound arriving from any direction. If the microphone responds to the velocity (speed and direction) of the sound wave, then it is a directional microphone, and the sensitivity has to be measured in terms of direction as well as amplitude of sound wave. The microphone types are known as pressure or velocity-operated, omnidirectional or in some form of directional response (such as cardioid).

The type of transducer does not necessarily determine the operating principle as velocity or pressure, because the acoustic construction of the microphone is usually a more important factor. If, for example, the microphone uses a sealed capsule construction, then the pressure of the sound wave will be the factor that determines the response. If the microphone uses a diaphragm or other moving element which is exposed to the sound wave on all sides, then the system will be velocity-operated.

The principle of the moving-iron or *variable-reluctance* microphone is illustrated in Figure 8.17. A powerful magnet contains a soft-iron armature in its magnetic circuit, and this armature is attached to a diaphragm. The magnetic

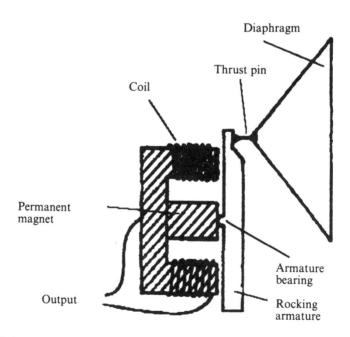

Figure 8.17 **Principles of a moving-iron microphone. The corresponding loudspeaker uses the same principles**

reluctance of the circuit alters as the armature moves, and this in turn alters the total magnetic flux in the magnetic circuit. This type of microphone has a high output level, with output impedance of, typically, several hundred ohms.

The moving-coil microphone, Figure 8.18, uses a constant-flux magnetic

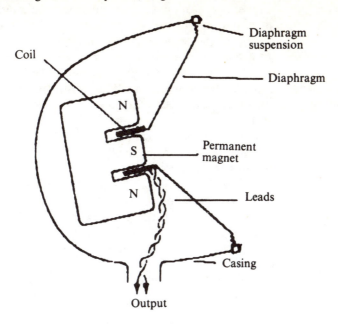

Figure 8.18   Principles of a moving-coil microphone. The corresponding
loudspeaker uses the same principles

circuit in which the electrical output is generated by moving a small coil of wire in the magnetic circuit. The coil is attached to a diaphragm, and the whole arrangement is usually in capsule form, making this pressure – rather than velocity-operated. The impedance is low, and the output is also low.

Other types of microphone include the ribbon type, in which a thin ribbon of corrugated foil is vibrated by the sound wave between the poles of a very strong magnet. The output is very low, as also is the impedance, and ribbon microphones are normally used along with a transformer or pre-amplifier. Piezoelectric microphones make use of the effect of sound waves on a vibration-sensitive crystal. Their output is high, with an impedance of 1M or more, and the linearity is poor. The capacitor microphone uses the principle that vibrating one plate of a capacitor will change the capacitance, so that if the capacitor is charged, the voltage across the plates will change in step with the vibration. Modern capacitor microphones make use of materials called *electrets*, which are permanently electrically charged, to permit the design of microphones with fairly high output and high output impedance. The electret material is the electrostatic equivalent of the magnet.

For each type of microphone, there is a corresponding earphone and loud-speaker type, whose basic principles are identical to those of the corresponding microphone. The conversion of electrical waves into sound waves is not sufficient, however, because the sound waves have to be launched into the air. This is comparatively simple in earphones because of the limited space in the ear, but loudspeakers have to be able to vibrate the air in an entire room, and this makes the type of enclosure (or cabinet) that is used of considerable importance.

The loudspeaker consists of a 'pressure unit', the transducer itself, and a diaphragm which will move a large volume of air. The majority of loudspeakers use the moving-coil type of transducer attached to a conical diaphragm. The diaphragm is one of the main problems of loudspeaker design, because it must be very stiff, very light and free of resonances, an impossible combination of virtues. The main problem of diaphragms is 'breakup', meaning that at high frequencies, different parts of the diaphragms will vibrate in different phases, distorting the sound. The usual solution to this problem is to use several loudspeakers in one enclosure, using a unit with a large cone (a woofer) for low frequencies and smaller units (tweeters) for high frequencies. The signals into the loudspeaker unit are separated by a form of filter called a 'cross-over unit'.

Moving-coil loudspeakers have a fairly low impedance, of the order of 3–4 ohms, and when mounted in an enclosure which is intended to smooth out the uneven response of the loudspeaker, the efficiency in terms of conversion of electrical power to acoustic (sound) power is very low, of the order of 1% or even less.

Though the moving-coil type of loudspeaker is dominant, ribbon and piezoelectric types are sometimes used for high-frequency signals. A few electrostatic loudspeakers (capacitor types) exist, and of those, the Quad electrostatic full-range types have acquired a formidable reputation in the twenty or so years for which they have been in production.

Multiple-choice test questions

1   A perfect transformer is defined by:
    (a) primary voltage = secondary voltage
    (b) primary current out = secondary current
    (c) primary power out = secondary power
    (d) primary turns = secondary turns.

2   For a transformer with a 5:1 stepdown ratio, a load of 1k in the secondary is equivalent to a load in the primary of:
(a) 5k
(b) 25k
(c) 200R
(d) 40R.

3   Which of the following light-sensitive transducers needs no power supply?
(a) phototransistor
(b) photoconductive cell
(c) photodiode
(d) photovoltaic cell.

4   A light meter uses a photodiode and an ordinary diode in the inputs to a balanced amplifier. The ordinary diode is used so as to:
(a) compensate for variations in temperature
(b) compensate for variations in light
(c) compensate for variations in power supply
(d) compensate for variations in amplification.

5   A thermistor is used in a temperature controller. You would expect the circuit to make use of:
(a) a video amplifier
(b) a balanced d.c. amplifier
(c) an audio amplifier
(d) an r.f. amplifier.

6   A loudspeaker system uses two moving-coil pressure units with cone diaphragms, one large, one small. Which of the following explains best the reason for this construction?
(a) This allows twice as much volume of sound to be generated.
(b) This allows each loudspeaker to handle the range of frequencies for which it is better suited.
(c) This allows the loudspeaker to convert more electrical energy into sound.
(d) This allows the enclosure to be a more convenient shape and size.

9   Digital circuits and displays

Summary

Digital signals. Binary code. Truth tables. Boolean algebra. Logic circuits. Gates. Combinational logic. Adders and comparators. Flip-flops. Clock circuits. Asynchronous and synchronous counters. Latches. Registers. Displays and decoders.

Note: Though the symbols for components such as transistors, resistors etc. are internationally agreed, the symbols for logical devices that are used for C&G and BTEC examinations in the UK are the British Standard (BS) set, which differ from the international set (of US origin). In this book, the international symbols have been used in the main because these are the symbols that will appear on virtually every data-sheet and circuit diagram that you are likely to encounter. You must, however, know the British Standard symbols for examination purposes.

Binary logic and arithmetic

The operation of *digital* circuits is controlled by two voltage levels only – either low voltage, referred to as *Logic 0*, or high voltage, referred to as *Logic 1*. The terms 'high' and 'low' are comparative only, since the 'high' voltage is often as little as $+5V$, and the 'low' about 0.2V.

By contrast, signals such as sine-waves (see Figure 9.1) constitute *analogue*

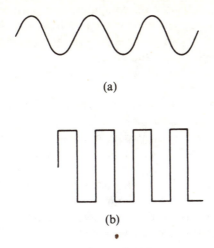

(a)

(b)

Figure 9.1   (a) Sine wave (analogue) and (b) square wave (digital)

signals. They have a continuously varying amplitude, lying anywhere between the peak values of the sine-wave voltage or current.

Two conventions define the levels of logic signals. These are shown in Figure 9.2. In *positive logic*, the most positive voltage level represents Logic 1. In

Positive and negative logic

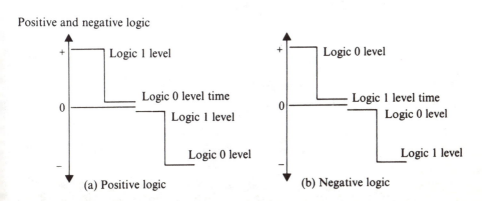

Figure 9.2   Positive and negative logic

231

negative logic, the reverse is the case, with the most negative voltage representing Logic 1.

All digital circuits used in small computers use positive logic, but some circuits used for industrial controllers make use of negative logic. A very large number of positive-logic circuits use a + 5V supply, with voltages from + 2.5V to + 5V representing logic level 1 and voltages of 0V to + 1V representing logic level 0. Some circuits use + 15V to represent logic level 1 and − 15V to represent logic level 0.

All digital counting and computing circuits use the figures 0 and 1 – and those figures only – in what is called the *binary counting scale*. Table 9.1 below shows how decimal numbers are converted into binary, and vice versa. For the purposes of machine control, Logic 1 represents switch-on, and Logic 0 switch-off. It is as simple as that.

The digits 0 and 1 can also be used for counting purposes. Any set of number is simply a method of counting, and our use of number in sets of ten (using digital 0 to 9) has developed because we have ten fingers and ten toes. Logic systems use the two digits 0 and 1 mainly because it is possible to create very reliable systems if only two sets of voltage levels are used to indicate digits. A binary number will therefore use only the digits 0 and 1. This means that a binary number will require more digits than its scale-of-ten (*denary*) counterpart. We can, however, use the same methods of positioning digits as we do for denary numbers. A number such as 529, for example, means 5 hundreds (100 =

Table 9.1   Denary-to-binary conversion

Write down the denary number you wish to convert. Say it is	1065	
Divide by two; write the result below and the remainder (0 or 1) at the side	532	1
Now divide the result by two again, once more placing the result below and the remainder at the side	266	0
	133	0
	66	1
	33	0
Go on doing this until the last possible figure has been divided, leaving zero and a remainder of 1. Thus:	16	1
	8	0
	4	0
	2	0
	1	0
	0	1

Now read off the remainders in order from the foot of the column upwards, and you get the binary number. In the example given, the binary equivalent of 1065 is 10000101001.

Table 9.2   Binary-to-denary conversion

Write out the binary number and set above each of its digits its proper *place number*, starting with place no. 0 at the right-hand digit of the binary number and working to the left from there. Thus in the example given:

Place no. 10 9 8 7 6 5 4 3 2 1 0
Binary no. 1 0 0 0 0 1 0 1 0 0 1

Then consult the table below, which allots to each place no. its own denary equivalent. (You will find the table quite easy to remember once you spot that every denary equivalent is exactly double its predecessor.) Mathematically, from place no. 1 leftwards all the denary equivalents are consecutive rising powers of two – 2^0 (which is 1), 2^1 (which is 2), 2^2 (which is 4), 2^2, 2^3, 2^4, 2^5 etc.

Place no.	Denary	Place no.	Denary
0	1	9	512
1	2	10	1 024
2	4	11	2 048
3	8	12	4 096
4	16	13	8 192
5	32	14	16 384
6	64	15	32 768
7	128	16	65 536
8	256	17	131 072

Next, look for the 1s in your binary number (ignoring all the 0s). Write down in another table the place nos of all these 1s and, in a separate column, their denary equivalents. Then add up the latter column, and you have the denary equivalent of your binary number.

Thus, in the example given, the important place nos are 0,3, 5 and 10. (All the other place nos lie over zeros in the binary number, and so don't count.) Construct the final table thus:

Place no.	Denary equivalent
0	1
3	8
5	32
10	1 024
	1 065

10×10, or 10^2) plus 2 tens plus 9 ones. When we use a binary scale, we are working with a base of two instead of ten, so that the number 110 means 1 four ($4 = 2 \times 2$ or 2^2) plus one 1 plus zero ones. This allows us to use the same methods of writing numbers and carrying out arithmetic as we use for the familiar denary scale numbers.

We can start by looking at how we convert number between the two main scales of binary (more precisely, 8–4–2–1 binary) and denary. Table 9.1 shows how denary numbers can be converted into binary, and Table 9.2 shows how binary number can be converted into denary numbers.

The advantages of using digital systems are as follows:

1 The voltage difference between the 0 and 1 levels can be made large enough to ensure freedom from interference.
2 The maximum error in a system can be kept as low as ± half a digit.
3 Semiconductors used in digital circuits are either cut-off or bottomed, except during the vary rapid changeover period. Power dissipation is therefore very low, and circuits can be made almost immune to variations in either temperature or component values.
4 The use of digital logic in machine control makes it easy to couple machines to computers.
5 Digital logic circuits are particularly easy to produce in IC form.
6 Conversion from digital to analogue ('D to A') or from analogue to digital ('A to D') is possible. Digital methods can therefore be used to replace analogue circuits in many applications.

Logic gates

Logic gates are circuits having several inputs and one output. The voltage level, either 0 or 1, at the output depends only on the inputs which are present at any given instant of time. Logic gates are designed in such a way that any desired combination of inputs can be made to produce an output 1, but that no other combination of input signals can produce an output 1. These logic gate circuits are often called *combinational* circuits because the output signal will always depend on what combination of signals exists at the inputs.

The desired action of any logic gate system is governed by its so-called *truth table* and *Boolean expression*. Give suitable circuitry, any combination is possible – say, a gate whose output is 1 only when the inputs are 1,0,1,0 (in which case it is called a *four-input gate*). A truth table is a list of all possible inputs to a gate circuit along with the output or outputs that will be obtained for each set of inputs. When a large number of inputs exists, the truth table will be very large, because the number of lines of a truth table is 2^n, where n is the number of inputs. For 6 inputs, for example, the truth table would use $2^6 = 64$ lines. A Boolean expression is a 'shorthand' mathematical way of writing the action of a logic circuit and is very much more compact once you know how to use the system.

Rather than produce an integrated circuit gate for every possible application, manufacturers make standardized circuits which can be used just like other circuit components. Some important standard gates, together with their International and British Standard symbols and truth tables, are shown in Figure 9.3.

The most important of the manufactured standard gates are the AND, NAND, OR and NOR types.

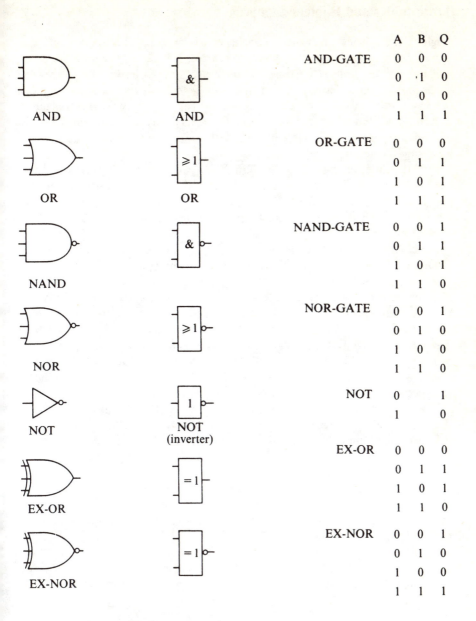

		A	B	Q
AND-GATE		0	0	0
		0	·1	0
		1	0	0
		1	1	1
OR-GATE		0	0	0
		0	1	1
		1	0	1
		1	1	1
NAND-GATE		0	0	1
		0	1	1
		1	0	1
		1	1	0
NOR-GATE		0	0	1
		0	1	0
		1	0	0
		1	1	0
NOT		0		1
		1		0
EX-OR		0	0	0
		0	1	1
		1	0	1
		1	1	0
EX-NOR		0	0	1
		0	1	0
		1	0	0
		1	1	1

Note: The circle indicates an *inverted* signal.

Figure 9.3   Logic symbols and truth tables

235

Truth tables and Boolean algebra

Logic is the science of drawing conclusions from facts and at one time was taught extensively. Logic, like electrical science, is based firmly on a few rules, and these rules, unlike the rules of electricity, have been known for several thousands of years. Throughout most of that time, however, the rules had been expressed in words and were not necessarily easy to apply. Boolean algebra, named after the mathematician George Boole, is a way of writing logical rules and relationships in abbreviated forms, and it is particularly useful as a way of analysing the action of gate circuits.

The truth value for the output Q of an AND gate with two inputs A and B is written Q = A.B. This means that the output is only at Logic 1 when both A and B are Logic 1. Similarly the output for a two-input OR gate would be written Q = A + B, the output being Logic 1 when either A or B is Logic 1.

Note that the symbols . and + have been chosen to represent AND and OR respectively, because they behave in logic in a way similar to that in which multiplication and addition behave in normal arithmetic.

When an input/output relationship is stated as being the *inverse* of the other, as in the case of the NOT gate, this is written $Q = \bar{A}$ – with the bar over the letter A representing 'NOT A'. Some manufacturers' data sheets now use # to mean NOT, so that A # means NOT-A and B # means NOT B, with (A + B) # meaning NOT (A OR B). This makes logical expressions easier to type and print. The method using the bar will, however, be used throughout this book.

The written expressions for NOT AND = NAND and NOT OR = NOR then become $Q = \overline{A.B}$ and $Q = \overline{A + B}$ respectively.

The two symbols labelled 'EX-OR' and 'EX-NOR' in Figure 9.3 stand for 'Exclusive OR' and 'Exclusive NOR' respectively. It will be seen that the 'Exclusive OR' function is similar to that of the OR gate except that it rules out the case where A = B = 1. The Boolean expression can be read from the truth table as $Q = \bar{A}.B + A.\bar{B}$

The 'Exclusive NOR' function is obviously the negation of 'Exclusive OR'. Its output is *true* (i.e. at Logic 1) when both A and B have the same value. It is sometimes called for this reason the *coincidence gate*. The Boolean expression for it is $Q = \bar{A}.\bar{B} + A.B$

A relationship between the 'Exclusive OR' and the 'Exclusive NOR' functions can be developed (see later), using the Boolean expressions and the NOT function as follows:

$$\bar{A}.B + A.\bar{B} = \overline{\bar{A}.\bar{B} + A.B}$$

Many other more complex logic functions can be equated in a similar way by using truth tables or the rules of Boolean algebra.

The ICs used in digital logic circuits

Two main families of IC circuits are used in digital circuits – the TTL (transistor-transistor-logic) and the CMOS (complementary-metal-oxide-silicon).

TTL ICs (see Figure 9.4) use bipolar transistors in integrated form, with their

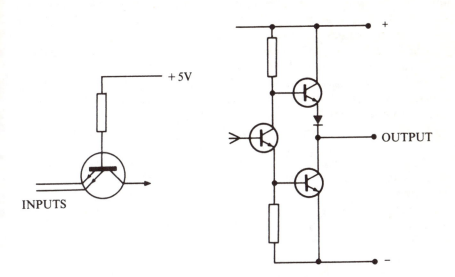

Figure 9.4   The TTL (transistor-transistor-logic) IC

input or inputs always connected to the emitter of a transistor whose base is connected, through a resistor, to +5V. With this type of logic, an input which is kept at Logic 1 (+5V) consumes no current (apart from a small leakage current); but an input kept at Logic 0 (zero volts) will pass 1.6mA, so that the connection between the gate input and zero volts must possess low resistance.

Each gate output can drive a current of, typically, 16mA either to earth (supplying current) or from the +5V line (sinking current). Each gate output can therefore drive up to ten other gate inputs (AND, OR, NAND, NOR, NOT types). In the language of logic design, the gate has a *fan-out of ten*. The opposite term, *fan-in*, refers to the number of possible inputs to a gate which need to be driven.

Some care has to be taken if more than one output is used to drive an input. The usual gate output circuit is a totem-pole circuit, in which the output terminal can be shorted through one of the transistors either to earth or to +5V. It is important that gates using this type of output circuit (as most do)

237

should never have their outputs connected together; for if one output were at 1 and the other at 0, destructively high currents would pass through both gates.

When gate outputs *must* be connected (in a type of circuit called the *wire-OR*), it is essential to use a gate with a 'floating' output. Such a gate has an output stage containing only one transistor, with no load. A resistor load external to the circuit itself must therefore be provided. It is known as a *pull-up resistor*.

An important point to note about all TLL circuits is that any unconnected input, or any input which has a resistance of more than 100 ohms in series, will 'float' to the base voltage of +5V because of emitter-follower action. At the high operating speeds normal for TTL gates, unconnected inputs can cause trouble through pickup of interference pulses. All unused inputs should therefore be connected to the +5V line through resistors of 1k or so.

For d.c. or low-speed operation, this precaution is less important.

In the CMOS family of ICs, field-effect transistors in their integrated form are used, with every input connected to an insulated gate. The input resistance is very high, so that input current is negligible whether the input signal is 1 or 0. Fan-outs of 50 or more are therefore possible.

A much wider range of supply voltages can be tolerated, though operating speeds are lower. Operating speed, is however, a significant factor only in large computers.

Careful circuit construction is necessary with CMOS devices because electrostatic voltages can cause breakdown of the gate of the FET (not of the logic gate itself) in the ways described when MOSFETs were being explained in Chapter 3.

Typical characteristics and operating conditions for TTL and CMOS gates are shown in the following table, which compares two gates performing identical logic functions.

Characteristics of TTL and CMOS digital ICs

General data	TTL	CMOS
Supply voltage	4.75 to 5.25V	4 to 12V
Fan-out	About 10	More than 50
Operating temperature	0°C to 70°C	−40°C to +80°C
4-input NAND gates		
Input current	40µA to 1.6 mA	10 pA
Output drive current	30 mA	0.25 mA
Switching time delay	11 ns	300 ns

Most IC gates are multiple, for it is as easy to package four two-input gates as one. The most commonly-encountered packages are the *DIL (dual-in-line)* 14- and 16-pin types whose pin-numbering schemes are shown in Figure 9.5.

Figure 9.6 shows a DIL package in outline.

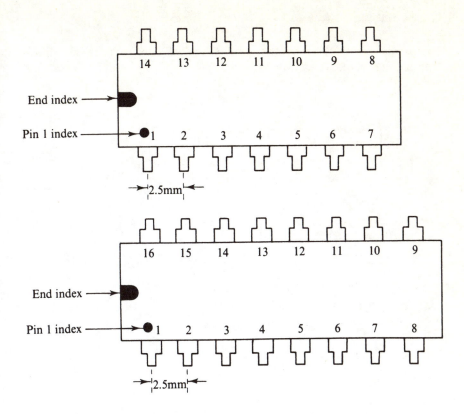

Figure 9.5   Dual-in line IC packages, 14-pin and 16-pin

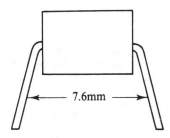

Figure 9.6   A DIL package in outline

239

Gate combinations

To establish the truth table for a combination of gates, such as that shown in Figure 9.7, proceed as follows:

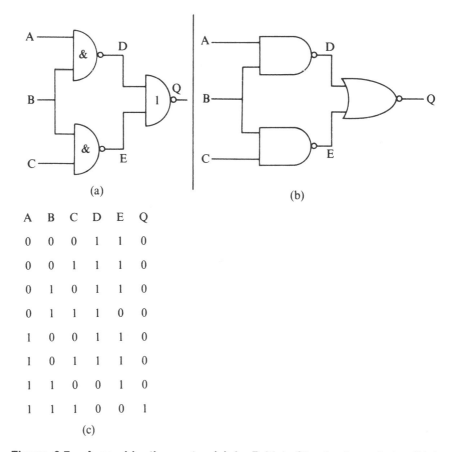

A	B	C	D	E	Q
0	0	0	1	1	0
0	0	1	1	1	0
0	1	0	1	1	0
0	1	1	1	0	0
1	0	0	1	1	0
1	0	1	1	1	0
1	1	0	0	1	0
1	1	1	0	0	1

(c)

Figure 9.7   A combination gate: (a) in British Standard symbols; (b) in International symbols; (c) the truth table for such a gate

1 Draw up a table, allowing one column for each gate input, and one for the final output.
2 Enter in the input columns for the first set of gates (cols A, B and C) every possible combination of 1 and 0. There will be 2^n such entries, where n is the number of inputs to the first set of gates. (In Figure 9.7 there are three inputs, and $2^3 = 8$). The entering is best done methodically by inserting in

the proper order the binary equivalents of the digital numbers 0, 1, 2 ... (in this case) 7. Thus the entries: 000, 001, 010, 011, 100, 101, 110, 111, in Figure 9.7 (c) make the 8 lines required for the three inputs to the system.

3   With the truth tables for every combination of gates known from Figure 9.3 enter the outputs of the first set of gates in cols D and E respectively. These outputs now constitute the inputs for the next set of gates, which are entered in col. Q.

4   Had there been more than eight inputs into the gate combination under analysis, the same procedure would have been followed until the table was complete and the final output signal determined.

Example. Find the truth table for the circuit of Figure 9.7.

Solution. Note first that Figures 9.7 (a) and (b) depict the same circuit, (a) showing it in British Standard notation and (b) in the International notation. The circuit consists of two NAND gates followed by a NOR gate. Turn to Figure 9.3 for the truth table relevant to a two-input logic gate so constructed.

Then draw up a blank truth table, labelling the system inputs A, B and C. Label the intermediate inputs D and E, and the output Q. Enter in the table all the possible values of inputs, using the ordinary binary sequence, 000, 001, 010, 011, 100, etc. This should automatically give the correct number of lines in the table.

Now enter col. D from the truth table for the NAND gate, which is always 1 except when both inputs are 1. Fill in col. E with the output of the other NAND gate.

With cols D and E complete, col. Q can also be completed – using this time the truth table for the NOR gate, an output which is 0 if either D or E is 1. The final truth table will look as illustrated in Figure 9.7 (c).

Exercise 9.1

For this and other exercises using TTL circuits, a suitable 'breadboard' for mounting and connecting TTL ICs is needed. The prototyping boards supplied by RS Components Ltd are ideally suited, and some types are complete with power supplies. ICs are plugged into the spaces provided, and other connections made by plugging components or wires into the holes on the board.

Connect a 7400 IC into place, using the circuits shown in Figure 9.8 (a and b). The numbers shown on the diagram correspond to the pin numbers of the 14-pin IC package. Using the input switches, and a voltmeter to give the output voltages, draw up the two truth tables. Note that unless an input is connected to zero, its value is 1.

A 5V supply must be used. Higher voltages can damage the ICs, while the use of lower voltages will cause uncertain action.

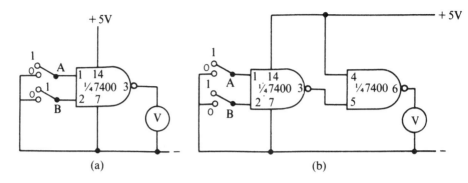

Figure 9.8   Circuits for Exercise 9.1

Fan-out and fan-in

Because logic gates are manufactured in standard IC forms, it is necessary to make logic circuits by connecting several standard logic gate units to each other, often in very large numbers. From the foregoing, it follows that any one logic gate output may have to drive several logic gate inputs. The ability to drive a number of inputs in parallel is described as the *fan-out* for a gate. Typically the fan-out is 10 for the TTL logic family. For the CMOS logic family, where the input drive current is very small, fan-out can be as high as 50. By comparison, the term *fan-in* refers to the number of inputs to a gate and is therefore independent of the logic family.

Combinational logic

Figure 9.9 (a) shows the truth tables for a selection of functions based on AND and OR gates. It proves an important logic identity known as De Morgan's theorem. This states that for a system with two inputs A and B, $\overline{A.B} = \overline{A} + \overline{B}$ and $\overline{A} + \overline{B} = \overline{A.B}$, this explains why the circuits in Figure 9.9 (b) perform the same logic function.

De Morgan's theorem can be further expanded for more than two inputs to show that:

$$\overline{A.B.C......} = \overline{A} + \overline{B} + \overline{C} + ... \quad \text{and}$$
$$\overline{A + B + C + ...} = \overline{A}.\overline{B}.\overline{C}.....$$

The theorem also explains why the same logic function can be generated using either NAND or NOR gates.

A	B	A.B	A + B	$\bar{A}$	$\bar{B}$	$\bar{A}.\bar{B}$	$\bar{A} + \bar{B}$	$\overline{A.B}$	$\overline{A + B}$
0	0	0	0	1	1	1	1	1	1
0	1	0	1	1	0	0	1	1	0
1	0	0	1	0	1	0	1	1	0
1	1	1	1	0	0	0	0	0	

(a)

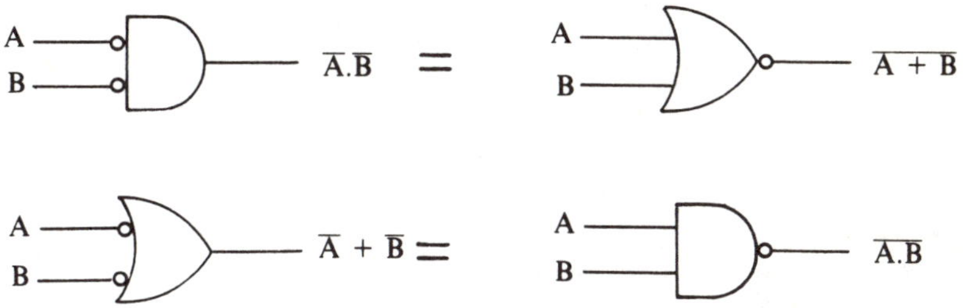

(b)

Figure 9.9 **(a) and (b) Truth tables and De Morgan's theorem**

Exclusive OR and coincidence detector

Figure 9.10 (a) and (b) shows the logic diagram and truth table for an Exclusive OR gate whose output excludes the possibility of A.B. The Boolean expression may be written, $F = A \oplus B$ where the symbol $\oplus$ stands for Exclusive OR. If the OR gate is replaced by a NOR gate, the truth table becomes that of Figure 9.10 (c), showing that the output is only logic 1 when A = B (a coincident gate). The Boolean expression now becomes $F = \overline{A \oplus B}$.

Boolean identities

The following *identities* (tables of equivalent quantities) can be proved easily by drawing the appropriate AND or OR gates and showing the inputs applied to them. The reader is invited to carry out these proofs as an exercise, using A = 0 and A = 1 and showing that each statement is true for either value of A.

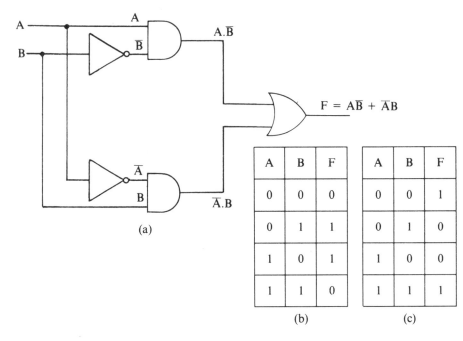

Figure 9.10 **(a) and (b) Exclusive OR circuit and truth table and (c) truth table for Exclusive NOR gate**

$A + 0 = A$
$A . 0 = 0$
$A + 1 = 1$
$A . 1 = A$

$A + A = A$
$A . A = A$
$A + \bar{A} = 1$
$A . \bar{A} = 0$

Adder circuits

The rules for binary addition can be stated as follows:

$0 + 0 = 0$
$0 + 1 = 1$
$1 + 0 = 1$
$1 + 1 = 0$ and carry 1

Thus two binary numbers can be added as follows, starting as usual with the digits (the least significant digits) on the right hand side and taking any carry digit one place left:

A	B	SUM	CARRY
0	0	0	0
0	1	1	0
1	0	1	0
1	1	0	1

(a)

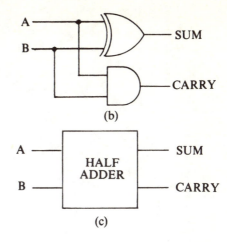

(b)

(c)

Figure 9.11 **(a) Rules of binary addition; (b) and (c) circuit and block diagram of half adder**

```
  01011010
  01101011
  ────────
  11000101
```

These rules are shown summarized in the truth table of Figure 9.11. From the sum and carry columns it can be seen that we need to combine Exclusive OR and AND gates to perform the operation of addition on two binary numbers A and B. The logic circuit is shown in Figure 9.11 (b). Because this circuit does not cater for a *carry in*, it is described as a *half adder*, and its block diagram symbol is shown in Figure 9.11 (c). A half adder is therefore only suitable for adding the least significant bits (LSB) of a pair of binary numbers. A circuit that does provide for 'carry in' is called a 'full adder' and its implementation and symbolic diagrams are shown in Figure 9.12 (a) and (b).

Exercise 9.2 – full adder

1 Using binary addition, complete the truth table.
2 Deduce the Boolean expression for both the sum S and the carry-out C_{out}.

Parallel adders

The addition of two binary numbers can be performed with a single full adder, producing the sum bit by bit in serial form, starting with the LSB first. This,

245

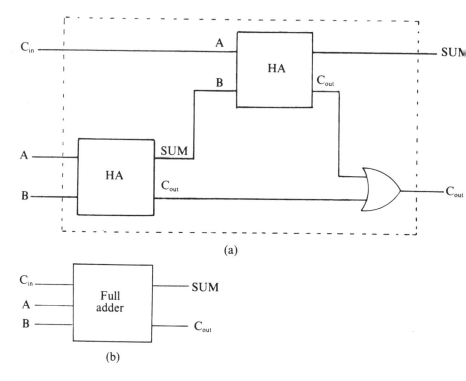

Figure 9.12   (a) Implementation of full adder and (b) symbolic diagram

however, is a relatively slow process. A parallel adder of the type shown in Figure 9.13 is much faster, but at the expense of a full adder per bit of binary number. All the bits of the number A are presented to the A inputs whilst the bits of the number B are simultaneously applied to the B inputs. Note, that whilst a half adder could be used for the LSB, a full adder is used in the interests of economy of component types. The carry-in of this stage is always zero because of the earth connection.

The comparator

It is sometimes necessary to know whether one binary number is greater than, equal to, or less than another binary number. This can be determined with a *magnitude digital (or binary) comparator*, often described as a *comparator* for short. However, care must be taken not to confuse this function with the analogue differential amplifier described in Chapter 4 which is also often loosely known as a comparator.

The Exclusive NOR gate has been shown to be an equality detector (output is

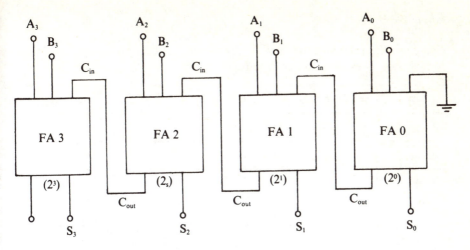

Figure 9.13   A parallel full adder

Logic 1 when A = B) and can thus form a part of the binary comparator. The Boolean expression for the Exclusive NOR gate is $F = \overline{\overline{A}.B + A.\overline{B}}$. Now if A = 1 and B = 0, A is greater than B, so that $A.\overline{B} = 1$. Also if A = 0 and B = 1, A is less than B, so that $\overline{A}.B = 1$. Figure 9.14 shows how these last two states can be met by extending the Exclusive NOR circuit with two AND gates.

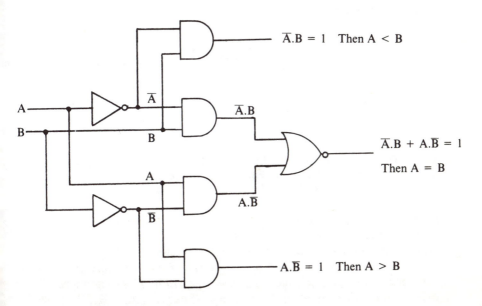

Figure 9.14   The comparator

247

Exercise 9.3

Complete the following truth table for Figure 9.14 to prove that the binary comparator gives the required results.

A	B	$\overline{A}$	$\overline{B}$	$\overline{A}.B$	$A.\overline{B}$	$\overline{A}.B + A.\overline{B}$	$\overline{\overline{A}.B + A.\overline{B}}$	Result
0	0	1	1	0	0	0	1	A = B
0	1							A < B
1	0							A > B
1	1							A = B

(Answers at end of chapter)

The terms *multiplex* and *demultiplex* often occur in digital work. Multiplexing means the use of a number of signal paths to take more information that could be expected, and demultiplexing refers to the reverse process. A simple example of multiplexing in this sense can be represented by signals for the numbers 0 to 7. If we used three lines for this purpose by representing the numbers in binary, these signals could be called multiplexed – three lines are used to carry signals for numbers 0 to 7, and the outputs cannot be directly used to represent these numbers (Figure 9.15). A demultiplexed form would use eight lines for the

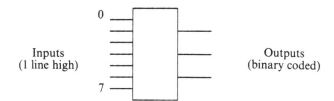

Figure 9.15 **An eight-to-three line multiplexer. The input is one '1' bit on one of the eight lines, and the output is a binary number corresponding to the input line number**

signals, and the outputs could, for example, each light a lamp whose position would indicate the value of the number.

Another meaning of multiplexing is more correctly described as *time multiplexing* in which a line or set of lines carries signals that represent different quantities at different times, such as one set of quantities while a timing pulse is positive and another set when the pulse is negative or zero. Figure 9.16 shows the form of the truth table for a 8 to 3 line multiplexer which will convert from

	INPUTS							OUTPUTS		
0	1	2	3	4	5	6	7	Q_0	Q_1	Q_2
1	0	0	0	0	0	0	0	0	0	0
0	1	0	0	0	0	0	0	1	0	0
0	0	1	0	0	0	0	0	0	1	0
0	0	0	1	0	0	0	0	1	1	0
0	0	0	0	1	0	0	0	0	0	1
0	0	0	0	0	1	0	0	1	0	1
0	0	0	0	0	0	1	0	0	1	1
0	0	0	0	0	0	0	1	1	1	1

$Q_0 = A1 + A3 + A5 + A7$
$Q_1 = A2 + A3 + A6 + A7$
$Q_2 = A4 + A5 + A6 + A7$

Figure 9.16   The truth table and equations for the simple multiplexer

denary input (a 1 on a line indicating the digit that the line represents) to binary output. From the truth table you can see what the requirements are for each 1 in the output columns. For example, the Q_0 output will be 1 when $A1 = 1$ or $A3 = 1$ or $A5 = 1$ or $A7 = 1$ but at no other times. In Boolean terms, this is:

$Q_0 = A1 + A3 + A5 + A7$

Similarly, we can write the Boolean expression for Q_1 which is:

$Q_1 = A2 + A3 + A6 + A7$

and for Q_2 which is:

$$Q_2 = A4 + A5 + A6 + A7$$

and this indicates that the task of multiplexing can be carried out with three OR gates, each with four inputs. The circuit is illustrated in Figure 9.17.

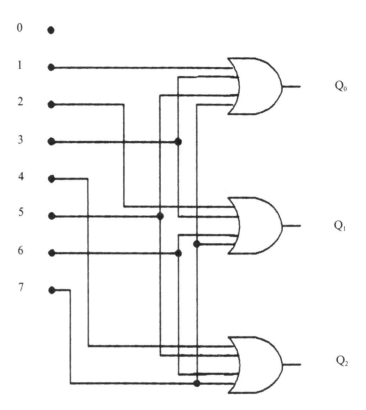

Figure 9.17   A circuit for a multiplexer which implements the truth table of Figure 9.16

The opposite section is demultiplexing, using the same truth table in the opposite direction. One possible demultiplexing circuit, for four binary digits in this case, is illustrated in Figure 9.18, using AND gates and inverters, and you should check through this in order to verify that it does indeed carry out the demultiplexing action.

The demultiplexer is a form of decoder, transforming a coded signal into one that can more easily be used or interpreted, but we often reserve the term *decoder* to mean a conversion from a number code other than binary. Suppose, for example, that we are using a display to show the progress of slow counting in binary numbers. The display consists of four bars arranged in a square, and the

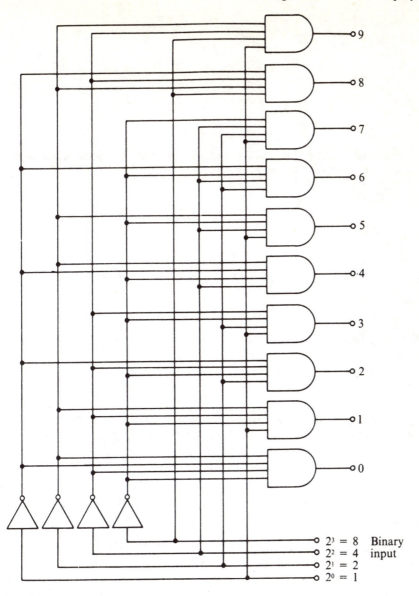

Figure 9.18

count sequence lights the bars in turn as indicated in Figure 9.19.

As before, the way to solve the problem is to draw up a truth table, Figure 9.20(a). This shows the values of Q_0 and Q_1 in the binary count that gives the values a to d for lighting the bars of the display, and from this we can draw up the Boolean expressions:

251

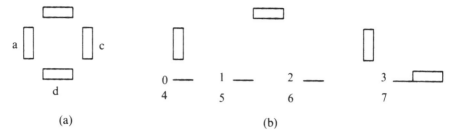

Figure 9.19   A set of bar LEDs and how they are intended to change with input numbers

$a = \overline{Q_1}.\overline{Q_0}$
$b = Q_0.\overline{Q_1}$
$c = Q_1.\overline{Q_0}$
$d = Q_0.Q_1$

pointing to the use of four AND gates and two inverters as shown in Figure 9.20(b).

R-S flip-flops

A flip-flop is a circuit of the bistable family. IC flip-flops tend to be of complex designs which would be uneconomical to manufacture with discrete components.

Figure 9.21 shows a very simple flip-flop circuit, the so-called R-S flip-flop which can be formed (among other ways) by using two NAND gates connected as shown. (R-S flip-flops are also, of course, available as ICs in their own right).

The R-S flip-flop has two signal inputs, labelled respectively R and S, and two outputs, labelled Q and $\overline{Q}$, with the bar indicating an inverse in the usual way. The truth table for this circuit demonstrates an important point distinguishing this and all other flip-flop circuits from simple gate circuits. Flip-flops belong to a class of circuits called *sequential logic circuits* in which the output Q has a value which depends on the previous value of the inputs as well as on their present value. The output of a logic gate, by contrast, depends only on the values of inputs which are present at the same time as the output. Thus in Figure 9.21 the truth table output for the inputs $R = 1$ and $S = 1$ depends on what values of R and S existed immediately before the $R = 1$, $S = 1$ state. If the previous state was $R = 0$, $S = 1$, then the arrival of $R = 1$, $S = 1$ gives $Q = 1$. If the previous state was $R = 1$, $S = 0$, then the arrival of $R = 1$, $S = 1$ gives $Q = 0$.

Note that there is no line $R = 0$, $S = 0$ for the R-S flip-flop. Such an input would give $Q = 1$, $\overline{Q} = 1$, which is not permissible. Moreover, R-S flip-flops can only be used when the state $R = 0$, $S = 0$ cannot possibly arise. For this reason

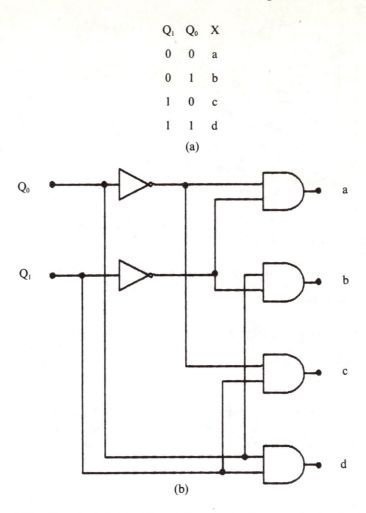

Q_1	Q_0	X
0	0	a
0	1	b
1	0	c
1	1	d

(a)

(b)

Figure 9.20   The truth table (a) and the circuit diagram (b) for the bar-LED driver logic

other types of flip-flops, such as the D-type and the J-K type, are used in most logic applications.

Exercise 9.4

Connect two gates of a 7400 in the form of an R-S flip-flop, as in Figure 9.21. Use voltmeters to indicate the states of Q and $\bar{Q}$, and switches to provide the inputs. Draw up the truth table.

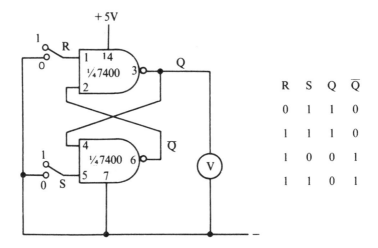

R	S	Q	$\overline{Q}$
0	1	1	0
1	1	1	0
1	0	0	1
1	1	0	1

Figure 9.21   The R-S flip-flop and its truth table

Clocked flip-flops

The simple R-S flip-flop changes state when either input becomes, or is set to, zero. Most digital circuits require flip-flops which change state only when told to do so by means of a pulse, called a *clock pulse*, arriving at a separate input. In such *clocked flip-flops*, all the flip-flops in a circuit can be made to change state at the same time.

Clocking has the advantage that the inputs need only be set an instant before the arrival of the clock pulse – changes of state at the inputs at any other time having no effect on the output. Between clock pulses, the output remains as it was when set by the last clock pulse.

D-type flip-flops

The so-called *D-type flip-flop* shown in Figure 9.22(a) triggers on the leading edge of the clock pulse. The logic value on the D input is then transferred to the Q output, thus delaying the data by the period of one clock pulse.

Reference to the truth table in Figure 9.22(b) shows the action of the *Preset* *(Pr)* and *Clear (Clr)* inputs. They are used to set the initial output states of the device. When Pr = 0, Q is set to 1. When Clr = 0, Q is set to 0. The states of both Clk and D are irrelevant at this moment of time because Pr and Clr override them. Only when Pr = Clr = 1 will the clock pulse transfer the data, and even then only as the clock pulse reaches a high value.

The state where Pr = Clr = 0 represents an unstable condition in a D-type flip-flop, and is prohibited.

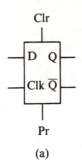

Pr	Clr	Clk	D	Q	Q̄
0	1	X	X	1	0
1	0	X	X	0	1
0	0	X	X	1	1
1	1	↑	1	1	0
1	1	↑	0	0	1
1	1	0	X	Q	Q̄

(a)   (b)

Note: X = any state; ↑ = goes high

Figure 9.22   The D-type flip-flop

J-K flip-flops

The most common type of clocked flip-flop is the J-K type represented by the symbol shown in Figure 9.23. This type of flip-flop has three signal inputs

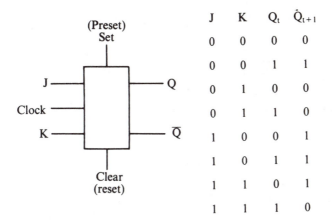

J	K	Q_t	$\dot{Q}_{t+1}$
0	0	0	0
0	0	1	1
0	1	0	0
0	1	1	0
1	0	0	1
1	0	1	1
1	1	0	1
1	1	1	0

Figure 9.23   The J-K flip-flop: symbol and truth table

labelled 'set' and 'reset' (or sometimes 'pre-set' and 'clear') which work independently of the clock pulses, and the usual two outputs Q and Q̄ (with Q̄ always the inverse of Q for any combination of inputs).

The J and K inputs control the voltage levels to which the output will change when the *trailing* edge of the clock pulse arrives. When the *leading* edge of the clock pulse arrives, the two inputs are used to pre-set what the output will

255

eventually be, but the change-over itself only takes place when the trailing edge of the clock pulse arrives.

The truth table for the J-K flip-flop is also shown in Figure 9.23, in which Q_t stands for the output voltage at Q before the clock pulse arrives and Q_{t+1} for the output at the Q terminal after the clock pulse has passed. The 'Set/Reset' inputs are used to change the value of Q at any time before, during or after the arrival of the clock pulse. Connecting the 'Set' terminal to zero volts causes the output at Q to go to Logic 1; connecting the 'Reset' terminal to zero volts causes the output at Q to go to Logic 0.

The familiar bistable counting action described in Chapter 7 occurs when both J and K inputs are taken to $+5V$ (in the case of TTL circuits). Two clock pulses arriving at the clock terminal give rise to one pulse at the Q output, while a pulse of inverse polarity will simultaneously reach the $\bar{Q}$ output.

Exercise 9.5

Connect the circuit shown in Figure 9.24 using a TTL 7476 IC, which contains

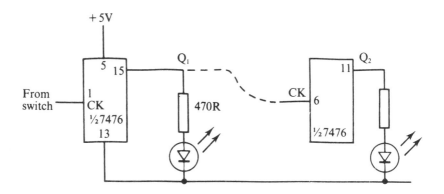

Figure 9.24   Circuit for Exercise 9.5

two J-K flip-flops, and preferably a breadboard. Apply clock pulses either from a slow pulse generator or from a 'debounced switch' (see below), and observe the output indicator LEDs.

Now connect the clock terminal (pin 6 of the second flip-flop in the package) to the Q-output (pin 15 of the first flip-flop), and re-apply the slow clock pulses. Observe the indicators.

Remember that all unconnected pins of TTL ICs will 'float' to $+5V$, so that no connections need be made to any pin which is to be kept at Logic 1 when such slow clock pulses are applied. When CMOS logic systems are used, no pin must ever be left disconnected.

Switch de-bouncing

When the output of a mechanical switch – either hand-operated or forming part of a thermostat, pressure switch, etc. – is used to feed pulses to a digital circuit, some precautions have to be taken to prevent 'bounce'.

When two mechanical contacts are made, the elasticity of the metals involved generally gives rise to one or more tiny bounces, each lasting a millisecond or so, before the switch contacts finally close. The waveform derived from such a switch closure is shown in Figure 9.25. If fed into the clock input of a counter,

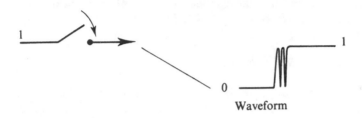

Waveform

Figure 9.25   The switch-bounce waveform

for example, it would cause serious miscounting, since each bounce would be counted as a switch operation. Circuits in which a mechanical switch is used to feed a counter must therefore have some means of eliminating the unwanted pulses caused by contact bounce.

Figure 9.26 shows two forms of de-bouncing circuit. The circuit of Figure 9.26(a) contains a two-way switch and an R-S flip-flop, and takes advantage of

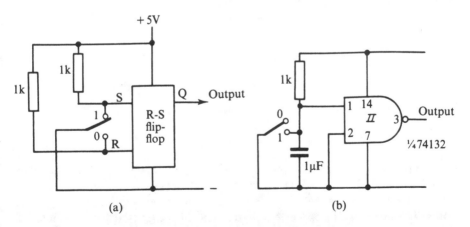

Figure 9.26   'De-bouncing' circuits: (a) with R-S flip-flop (b) with Schmitt trigger

the fact that the output of the R-S flip-flop does not change when one of its inputs is taken to Logic 1.

The current of Figure 9.26(b) employs a Schmitt trigger, in the form of a TTL IC gate with Schmitt characteristics, and an integrator at the input. If the switch contacts bounce, the capacitor cannot charge quickly enough to change the state of the trigger.

Note, by the way, that clock pulse inputs to digital circuits must have short rise and fall times. The inputs to gates also should never be slow-changing waveforms. The reason is that the gain of these circuits, considered as amplifiers, is very large, so that a slow-changing waveform at the inputs can momentarily bias the circuit in a linear mode, so making oscillation possible. Oscillation also can cause mis-counting and erratic operation.

Clock circuits (oscillators)

Sequential logic circuits are oscillator- or clock-driven. These may be *free-running* or synchronized to a data stream. Since clock frequencies are commonly in the megahertz range, a high degree of frequency stability is necessary. To this end, crystal-controlled clocks are common. In digital systems the active element of the clock circuit is usually a digital integrated circuit. The arrangements of Figure 9.17 and 9.28 are representative of this type of circuit.

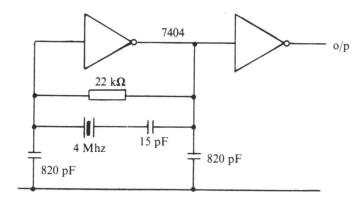

Figure 9.27 4MHz crystal-controlled clock

The circuit of Figure 9.27 uses one inverter section of a multiple inverter chip as the amplifier. In the analogue sense, a signal inversion is equivalent to 180° of phase shift. The crystal behaves as a tuned circuit of very high Q factor, with its opposite ends being of opposite polarity. This of course is also equivalent to a further 180° of phase shift. The resulting positive feedback produces oscillations at the crystal frequency. A second inverter acts as a buffer amplifier.

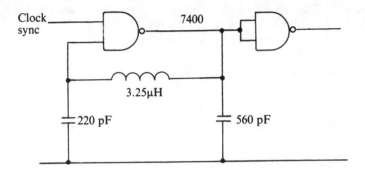

Figure 9.28   7MHz LC oscillator

It will be recalled that the output from a NAND gate is Logic 0 when its inputs are all at Logic 1. Again, in the analogue sense, this represents 180° of phase shift. The tuned circuit consists of L_1 in parallel, with C_1 in series with C_2. Opposite ends of L_1 are of opposite polarity so that again positive feedback is generated to produce oscillations so long as the sync input is positive. In this way the oscillations will be locked to a positive-going data stream. The second NAND gate functions as a buffer amplifier. The circuit of Figure 9.28 can be modified to use crystal control for better frequency/temperature stability.

Counting and decoding

A fundamental feature of all bistable devices is that they divide the input by two. Thus for every two clock-pulse inputs, the Q output changes once. This concept forms the basis of electronic binary counters, an example of which is shown in Figure 9.29.

The pulse sequence to be counted is the input to FF1, whose Q output now provides the clock for the next stage, and so on. A count input to the first stage

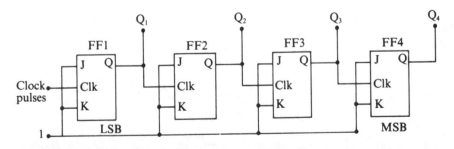

Figure 9.29   The ripple or asynchronous counter

259

thus 'ripples' through the circuit – which is why a counter of this type is often called a *ripple*, a *ripple-through* or an *asynchronous counter*. Note that the output count should be read from right to left, because the most significant bit (MSB) is on the right.

The ripple counter's waveforms are shown in Figure 9.30. Since J = K = 1,

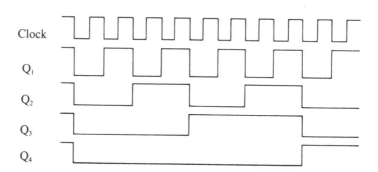

Figure 9.30   Binary counter waveforms

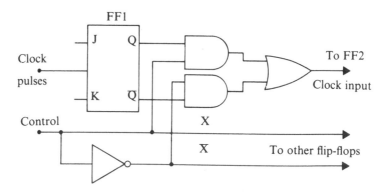

Figure 9.31   The up–down counter control

the Q output toggles, or changes, on the edge of every negative-going waveform.

When the Q̄ output of every stage of a flip-flop is used to provide the clock pulses, the circuit counts down from a preset number, to produce a *ripple-down* counter.

If the control circuit shown in Figure 9.31 is added between each flip-flop, it is possible to produce a counter which will count up or down according to the state of the control line. If control is Logic 1, the circuit will count up; while if the control is Logic 0, it will count down.

Synchronous counters

These ripple or synchronous counters suffer from one serious disadvantage – they are relatively slow. The *propagation delay* is the time required for the counter to respond to an input pulse. In the worst case, when all the flip-flops are at logic 1, the next pulse must reset all the flip-flops to zero. This input pulse thus ripples through the circuit. If the total propagation delay is longer than the periodic time of the pulse stream, some pulses simply do not get counted. If the operation is changed so that all the flip-flops are clocked at the same time (synchronously) the propagation delay is reduced to that of a single flip-flop thus increasing the maximum count rate. For a 4-bit counter, the maximum count frequency might rise to above 50MHz compared with about 25MHz for a ripple counter.

Divide by N counters

It may be desired to divide by a number N which is not a power of 2; for instance, divide by 10 for a decimal counter. To construct such a counter n flip-flops are needed, such that n is the smallest number for which $2^n > N$. A feedback circuit detects when a count of N has been reached so that the circuit can be reset to zero. For a decimal counter $n = 4(2^4 = 16 > 10)$. Decimal 10 is 1010 in binary, so that the feedback must detect this pattern to reset the counter to zero. An asynchronous or ripple decade counter is shown in Figure 9.32 with the least significant bit (LSB) on the left at FF0. The output count appears on Q_0 to Q_3. This needs to be decoded to display the count in decimal form. As soon as $Q_1 = Q_3 = 1$, the Clear (Cr) inputs immediately reset all the Qs to zero to restart the count. Note that Q_1 and Q_3 first become 1 after the tenth pulse and then are quickly reset to zero. This generates a very narrow pulse at the output which requires additional circuitry to suppress it. Ideally Q_1 and Q_3 should go to zero immediately on the tenth pulse.

A synchronous decade counter is shown in figure 9.33. Since the bistables are all clocked in parallel, it works much faster than the ripple counter. The J-K flip-flop only toggles or changes state when $J = K = 1$. The toggle action in this circuit being controlled via the Q outputs and the AND gates. Before any bistable can toggle, all the earlier bistable Qs must be at Logic 1. Because $J_0 = K_0 = 1$, FF0 toggles on every clock pulse. Whilst $Q_3 = 0, \bar{Q}_3 = 1$ and when $Q_0 = 1$, $J_1 = K_1 = 1$ and FF1 toggles. In a similar way, FF2 toggles when $Q_0 = Q_1 = Q_2 = 1$. When $Q_3 = 1$, $\bar{Q}_3 = 0$ so that $J_3 = 1$ and $K_3 = 0$ because

$Q_0 = 0$ on the tenth pulse (see waveforms of Figure 9.32). But at this time all the flip-flops reset to zero.

Applications of counters

Apart from direct counting, these devices can perform many functions in industrial process control. Objects on a conveyor belt can be counted using a photo-electric cell and a light source. The reset pulse can be used to initiate an action to close a bottle after it has been loaded with the preset number of pills. A downward count to zero is also useful. This is used on some automatic coil-

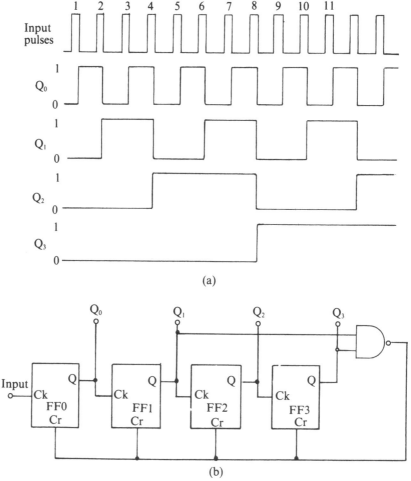

(a)

(b)

Figure 9.32 **(a) An asynchronous or ripple decade counter and (b) waveforms**

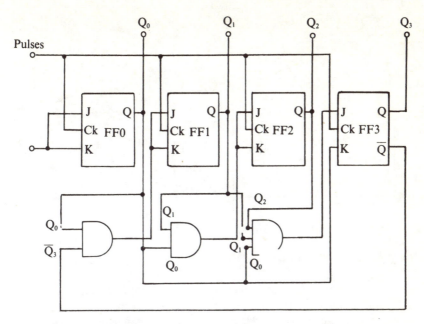

Figure 9.33   A synchronous decade counter

winding machines. The required number of turns to be wound on to a bobbin is first preset, and the machine set in motion. When the count reaches zero the machine automatically comes to a halt.

A counter can be used to measure frequency of both unipolar (d.c.) and bipolar (a.c.) waveforms by counting the number of cycles over a precisely controlled period of time. It can also be used to measure time if driven by a precisely known crystal frequency. This is the basis of the digital watch. In radar or sonar systems a transmitted pulse is reflected from an object and some of the energy returned to the source. The time delay can be measured and since the velocity of radio or sound waves is known, the time can be converted into distance.

Two photo-cell/light source pairs can be set up a known fixed distance apart. The time taken for an object to pass between them can be measured and the distance/time computed to give the velocity.

Latches

A latch is simply a 1-bit storage cell; a typical circuit is shown in Figure 9.34. The Q output takes up the state of the data input at each clock pulse and this is held until the data changes. The circuit has basically a bistable action. Apart from acting as a temporary data store, the latch can function in a similar way to the Schmitt trigger and can be used to de-bounce switch contacts. A number of

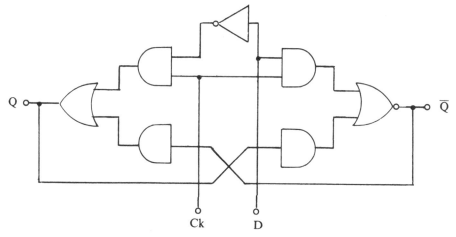

Figure 9.34   A latch circuit

latches can be formed on a single IC so that a complete binary word may be held. A particularly useful function for such a device is the address latch used on a computer memory matrix. Having decided on the particular memory cell required, its address has to be held long enough for the data to be written in or read out.

Shift registers

A register is formed by interconnecting a number of bistables in a similar way to a counter. Each bistable forms a memory cell for one bit of information. The data may be input in serial or parallel form and read out in a similar manner, depending on the method of interconnection. There are thus four basic forms of shift register:

1   Serial in–serial out (SISO).
2   Serial in–parallel out (SIPO).
3   Parallel in–parallel out (PIPO).
4   Parallel in–serial out (PISO).

Some registers are of one form only whilst others are more universal, the mode of data transfers then being controlled by the logic values of separate control lines. The general principle is explained using Figure 9.35, where the register is formed from D-type flip-flops (R-S flip-flops may also be used). It will be recalled that the logic value on the D input is transferred to the Q output on the next clock pulse. In this example, each data value input thus ripples through the register in a manner similar to the counter.

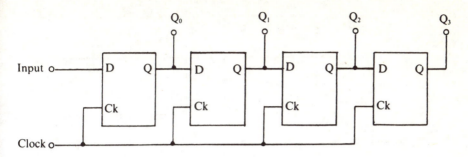

Figure 9.35   A 4-bit shift register

Exercise 9.6

The current state of the logic values for the shift register of Figure 9.35 is as follows:

$Q_0 = 1$
$Q_1 = 1$
$Q_2 = 0$
$Q_3 = 1$
$D = 0$

What is the state of the Q outputs after the next clock pulse? (*Answer at end of chapter.*)

The functions performed by shift registers include:

1 Multiplication and division (SISO).
2 Delay line (SISO).
3 Temporary data store or buffer (PIPO).
4 Data conversion; serial to parallel and vice versa (SIPO and PISO).

Serial input–parallel output shift register

Figure 9.36 shows a serial in–parallel out shift register that may be used to convert data from serial to parallel forms, as well as to perform multiplication and division, depending on the logic values on C_1 and C_2. With $C_1 = C_2 = 0$ the clock is inhibited so that no movement of data occurs within the shift register. When $C_1 = 0$ and $C_2 = 1$, the register is in the shift right mode so that data may be entered from the left-hand cell, least significant bit first. When $C_1 = C_2 = 1$, the data in each cell is loaded into the output latch circuit for a read out operation. When $C_1 = 1$ and $C_2 = 0$, the data held in the shift register can be left shifted on successive clock pulses to provide binary multiplication.

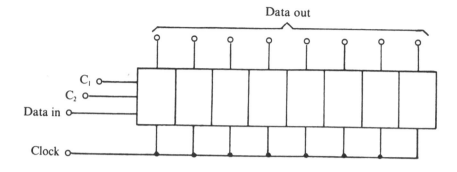

Inputs		
C_1	C_2	Mode
0	0	Hold
0	1	Shift right
1	0	Shift left
1	1	Parallel

Control Truth Table

Figure 9.36   A serial in–parallel out shift register with control truth table

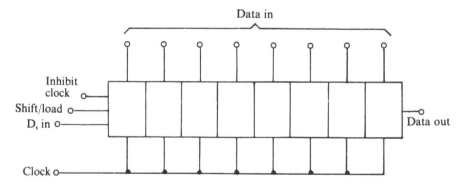

Figure 9.37   A serial/parallel input–serial output shift register

Serial or parallel in–serial out shift register

The example shown in Figure 9.37 shifts data to the right on each clock pulse. When the shift/load input is at logic 1, the serial input is inhibited and on the first clock pulse the parallel data is entered into each cell. When shift/load is switched to logic 0, the data is clocked out in serial form, one bit at a time. For

266

serial in–serial out operation the shift/load input is held at Logic 0 and data is entered through $D_s$in.

Displays

A binary number can be displayed (though they seldom are) by using filament lamps or LED indicators connected to the Q output of each bistable. Decimal displays, however, are much more useful, and the decoder system outlined above nearly always requires a decimal display to make its results visible.

Among the types of decimal display commonly used are incandescent (filament) lamps, cold-cathode displays, and LED and liquid crystal displays (LCDs). The techniques necessary to obtain a display differ somewhat from one type of display to another.

Incandescent or filament displays are commonly used to give very large displays – needed in some industrial applications and consisting of a set of lamps which illuminate slides on which decimal numbers are outlined. An optical projection system then displays the selected figure on a screen. Since each lamp displays one figure only, a binary-to-decimal decoder of the type mentioned earlier will be suitable, though a driver stage will generally be needed to supply the power needed by the lamps. Such a driver stage is often referred to as an *interface* – a general term use for any circuit needed to couple digital logic either to other circuits or to digital circuits operating at different voltage levels.

Cold-cathode displays rely on the fact that a gas at low pressure will conduct at a voltage between 45V and 210V, depending on the composition and pressure of the gas. When the gas conducts, a bright glow surrounds the cathode (negative) electrode which is in contact with the gas. If this cathode is made in the shape of a figure or a letter, the glow too will take this shape when the gas conducts. A cold-cathode display therefore comprises an anode (a nickel wire) and ten cathodes shaped in the form of the figures 0 to 9, all enclosed in a glass bulb containing a gas – usually neon – at low pressure. When a suitable voltage exists between the anode and one of the cathodes, a glow will appear around that cathode, so illuminating its figure.

Because of the high voltage levels involved, an interface circuit is also needed. For decoding, a binary-denary decoder can be used.

LED displays make use of the glow which occurs when current passes through a diode made of gallium arsenide, or of similar materials such as indium phosphide. Liquid-crystal displays (LCD) do not generate light, and must be viewed either by reflected light or by a light placed behind the display (a *backlight*). A liquid-crystal display makes use of a material which is placed between conducting plates, forming a capacitor, and which polarizes light when an electric field is applied between the plates. If the light which passes through the display is already polarized (by using a film of polaroid material) the

267

additional polarization of the LCD cell can prevent light from passing through, so that the activated display looks dark.

A typical reflective LCD cell uses a reflective metal film as the backplate and a transparent tin-oxide coating as the front plate, both films being deposited on glass. The LCD material itself is a liquid, of which cholesterol is one example, whose molecules are very long and which carry opposing charges at the ends. When a field is applied between the plates, the molecules turn to line up with the field, and this has the effect of polarizing the light that passes through the cell. For a reflection cell, a polaroid sheet is placed over the transparent side of the display, and when backlighting is used a polaroid sheet is placed over the rear window which for a backlit cell must also be transparent.

In practice d.c. must never be applied to the LCD display, because it causes the molecules to separate. The displays are operated by using a.c. fields, preferably at frequencies of several kHz. The voltage is typically around 30V, and since each LCD cell is a capacitor of very small value, the current that is required to activate a cell is very small, a few μA at the most even for a large cell. A backlit cell, of course, will require additional power to operate the backlight, but this need not be supplied from the signal drivers.

In either type of display, the materials are formed into the seven-segment shape shown in Figure 9.38 – the segments identified by letters as shown. From

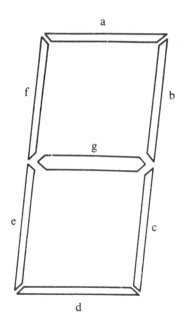

Figure 9.38   The structure of a seven-segement display

this basic shape, a complete set of figures (plus a few letters) can be constructed by selecting suitable segments for illumination.

A binary-to-denary decoder is unsuitable for this type of display, which requires a new kind of truth table for a binary-decimal seven-segment decoder. Such a truth table is shown in Figure 9.39.

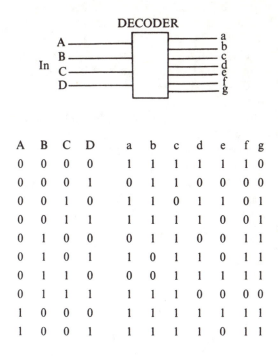

A	B	C	D		a	b	c	d	e	f	g
0	0	0	0		1	1	1	1	1	1	0
0	0	0	1		0	1	1	0	0	0	0
0	0	1	0		1	1	0	1	1	0	1
0	0	1	1		1	1	1	1	0	0	1
0	1	0	0		0	1	1	0	0	1	1
0	1	0	1		1	0	1	1	0	1	1
0	1	1	0		0	0	1	1	1	1	1
0	1	1	1		1	1	1	0	0	0	0
1	0	0	0		1	1	1	1	1	1	1
1	0	0	1		1	1	1	1	0	1	1

Figure 9.39   Truth table for a binary seven-segment decoder

LED and LCD forms of display account for most of the displays used for calculators and instruments, and their relative merits and drawbacks are tabulated below. LCD displays are used extensively for portable computers, and suitable types have now been developed for colour TV displays, particularly for small portable receivers.

LED advantages	LED disadvantages
Low voltage operation	Large current required
Clear bright display	High power consumption
Brightness easily controllable	Little choice of colour
Can be operated with d.c.	Destroyed by reverse voltage

269

LCD advantages	LCD disadvantages
Display easy to read in bright light	Needs backlight in darkness
Very low power consumption	Needs high-frequency drive
Colour displays possible	Small viewing angle
Very complex shapes possible	Difficult to control brightness
Each cell is a capacitor	D.c. must never be applied

Displays of very intricate shapes can be made by using several segments of LCD or LED units, but it is much easier to fabricate complex shapes in LCD then in LED, since the LED is basically a diode. The seven-segment display is an example of multi-segment use, but for more elaborate patterns the use of LCD is easier, though complex LED displays have been made using a matrix of dots with each dot independently controlled.

Multi-segment displays using LEDs can be manufactured as common anode or common cathode types. A common anode display has all of the diode anodes internally connected to one pin, taken externally to a positive voltage, and each segment is activated by driving its cathode, through a separate pin, to a lower voltage. The common cathode display has all of the segment cathodes taken to one pin which is usually earthed, and the display drivers raise the voltage of each anode which is to be activated. LCDs are basically capacitors, so that there is no such distinction required, one plate of each segment will be common and the other will be driven by an a.c. voltage when that segment is to be activated.

Either type of display will require a decoder, and for LCD types the decoder usually incorporates the d.c.–a.c. inverter circuits that provide the a.c. signals at high frequencies and voltages of typically 10V to 30V peak-to-peak.

Strobing

Many displays are *strobed*, which means that only one display at a time is switched on in a multi-digit display. This technique can be used to reduce the number of active circuits – though at the expense of added complexity (of circuit diagram, be it said, rather than of actual connections).

In a strobed display, one decoder driver is connected to all of the displays; and the counters are connected to the decoder-driver by a circuit called a *multiplexer*, which is a switch circuit operated by a coded input. A clock pulse, called in this operation a *strobe pulse*, is used to switch the multiplexer, and also to brighten up each display. The first figure from the counter is decoded and displayed on its segments; then the next strobe pulse switches the inputs of the decoder to the next stage of the counter – and the bright-up pulse to the next set of segments – so that the next figure is shown.

The strobing is carried out at very high speed, so that what the observer sees appears to be a continuous set of figures, rather than figures illuminated in very rapid sequence. Strobing circuits are often included as part of a large IC (such as

270

the clock ICs used for digital clocks), so that only the connections to the display require to be made.

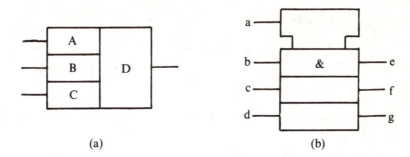

(a)   (b)

Note: These conventions apply to the British Standard symbols as used in C & G examinations. They are unlikely to be encountered in diagrams used by manufacturers of equipment.

Figure 9.40   Embedded or Combined logic circuit diagrams

New logic symbols

Figure 9.40 shows how the various logic elements in a single chip are combined in a circuit diagram. In Figure 9.40(a) the signal flow crosses only the vertical boundary so that there is no signal connection between A, B and C, but all three of these inputs feed output D.

The element shown at the top of Figure 9.40(b) is described as a common control block or element. With the AND gate combination shown here, the

outputs will be as follows:$E = A.B$ $F = A.C$. $G = A.D$

The triangular additions to the input/output lines shown in Figure 9.41(a)

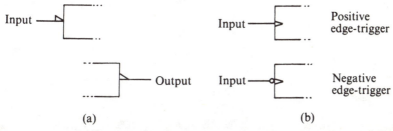

(a)   (b)

Figure 9.41   Other changes to British Standards (a) polarity or negative-logic indicators, (b) edge-triggering symbols

271

indicate that negative logic applies at these points. The symbols of Figure 9.41(b) are used to indicate the polarity of the trigger signal in flip-flop circuits.

More advanced topics

Note: The following material is required for the BTEC Electronics Unit B86/331 and may be useful to students of C&G 224 who are likely to progress to Part 3 studies relating to digital equipment.

The circuit in Figure 9.42 consists of two OR gates and an AND gate, and is

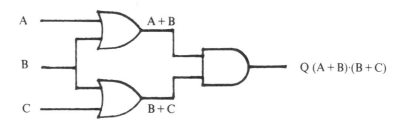

Figure 9.42   A gate circuit as an exercise in analysis

suspected to be more than is required to carry out the action. We start an analysis by writing the Boolean equation, which is:

$$Q = (A + B).(B + C)$$

and then we can 'multiply out' this expression, to give:

$$A.B + A.C + B.B + B.C$$

and we can start to simplify and then gather the terms that we have here. To start with, B.B simplifies to B, since 0 AND 0 is 0 and 1 AND 1 is 1. We can then regroup (the equivalent of factorization in ordinary algebra) in the form:

$$A.B + A.C + B.(1 + C)$$

and the term $(1 + C)$ can go out, because 1 or C must simply be 1. We are now left with:

$$A.B + A.C + B$$

and we can regroup this as:

$$A.C + B.(1 + A)$$

272

and once again remove the 1 OR A term, leaving:

B + A.C

so that the circuit action could be carried out by one AND gate and one OR gate, as in Figure 9.43. You can draw up the truth tables for these two circuits to

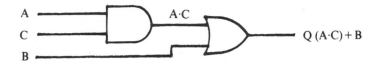

Figure 9.43   The simplified version of the circuit

check that they are identical.

Karnaugh mapping

Not everyone has a taste for the sort of algebraic manipulation that this first example uses, and there is an alternative method, called Karnaugh mapping, which can be used to simplify Boolean expressions. Karnaugh mapping involves drawing a form of short truth table from which redundant parts can be eliminated by using a set of simple rules. It looks rather like a game of noughts and crosses in action, and for anyone who is perplexed by the groupings and eliminations of the algebraic method it can offer an alternative that can be easier for gate circuits with not too many inputs. A Karnaugh map for a circuit with three inputs is relatively simple, for four inputs it is more difficult, and for more than four inputs you need hot tea and aspirin.

The principle is to draw up a form of truth table which will have a box for each possible term in the Boolean expression. These boxes are arranged in rows and columns, and for a three-input circuit, there are four rows (one for each possible state of two inputs) and two columns (one for each state of the remaining input). A blank Karnaugh map or three inputs therefore looks as in Figure 9.44. What you have to watch carefully is the order of the AB quantities. This does not follow a binary count; its order is 00, 01, 11, 10 reading downwards, a Gray-code sequence in which only one bit changes at each step. Depart from this order, and you will find yourself in trouble! To make use of this table, you need the expression in its fully 'multiplied out' version, consisting of a set of quantities that are read together. In the example we looked at, this is:

A.B + A.C + B.B + B.C

and we can use this to give a 1 or set of 1s for each term, so placing 4 1s into the

C

A B	0	1
00		
01		
11		
10		

Figure 9.44   A blank Karnaugh map for a circuit with three inputs

Karnaugh map. The first term is A.B, meaning that for the position where A = 1 and B = 1, we can place a 1. Since C does not appear in all this; we can place a 1 in both columns for the A = 1, B = 1 line (the 11 line), as shown in Figure 9.45.

C

A B	0	1
00		
01		
11	1	1
10		

Figure 9.45   Placing the 1s into the Karnaugh map to represent the A.B term in the Boolean equation

This is done for each term in the equation. The A.C line will place a 1 in the C = 1 column and the A = 1 rows, and since there is a 1 in the A = 1, B = 1, C = 1 box already we do not need to put in another. The B.B expression puts a 1 for each box where B is 1, regardless of the values of A or C – once again, some of these boxes are already filled in. The B.C term then puts 1s in where B = 1 and C = 1, and we find that these boxes are already filled. When you find that there is overlapping like this, it's a sure indication that there is quite a lot of redundancy in your circuit. For each box which does not contain a 1, you can now place a 0, Figure 9.46, though this is not strictly necessary.

The next crucial stage is the selection of the wanted parts of the map which are, in this case, the boxes that contain 1s. What we are interested in is a line or box of 1s, and in this example we can find three lines of them, as illustrated in

A B	C 0	1
00	0	0
01	1	1
11	1	1
10	0	1

Figure 9.46 Completing the Karnaugh map with 1s and 0s

A B	C 0	1
00	0	0
01	1	1
11	1	1
10	0	1

B   Square covers area where A = 0 or 1 and C = 0 or 1

A.C   Rectangle covers area where A and C are 1.

Expression is B + A.C

Figure 9.47 Marking groups of 1s on the map, and finding the logic terms that correspond to the groupings

Figure 9.47. The biggest number we have collected together and unbroken by any 0s is the set of four in the A = 0, B = 1 and the A = 1, B = 1 rows. The common factor here is B = 1, so this line represents a term B. The odd 1 out can be linked to the one above it, true of C = 1 and A = 1, so that it represents A.C. This makes the simplest expression B + A.C, as expected.

Karnaugh mapping, along with the use of De Morgan's theorem, can greatly simplify *minimization*, the reduction of a Boolean expression for a circuit to its simplest form. It is important to remember, however, that there is quite often no single simplest form, but several forms that on paper are equally simple. In practice, one form may be considerably simpler to implement because it can use standard units like quad NOR gates.

275

Multiple-choice test questions

1 A circuit is digital if it uses:
   (a) square waves
   (b) two levels of voltage only
   (c) frequency division
   (d) a $+5V$ supply.

2 A circuit is called combinational if:
   (a) its inputs are combined to produce the output
   (b) its output is a set of combined signals
   (c) the output is determined by the combination of inputs
   (d) each input is a combination of signals.

3 The action of a 4-input gate circuit can be most concisely expressed using:
   (a) Boolean algebra
   (b) truth tables
   (c) a description in words
   (d) a circuit diagram.

4 A sequential circuit can best be described as
   (a) a circuit containing flip-flops
   (b) a circuit containing gates
   (c) a circuit in which the inputs are in sequence
   (d) a circuit in which the output is determined by the sequence of inputs.

5 The main disadvantage of the asynchronous or ripple type of counter is that:
   (a) it needs to be constructed from J-K flip-flops
   (b) the last flip-flop cannot change state until all the preceding flip-flops have changed
   (c) the pulse to be counted is applied to all of the flip-flops
   (d) it needs to be constructed from D-type flip-flops.

6 A shift-register cannot be used for:
   (a) counting input pulses
   (b) storing input signals
   (c) delaying input signals
   (d) combining input signals.

Answers to exercises

9.2 $S = \bar{A}.\bar{B}.C + \bar{A}.B.\bar{C} + A.\bar{B}.\bar{C} + A.B.C$
     $C_0 = \bar{A}.B.C + A.\bar{B}.C + A.B.\bar{C} + A.B.C$

9.3

A	B	$\overline{A}$	$\overline{B}$	$\overline{A}.B$	$A.\overline{B}$	$\overline{A}.B + A.\overline{B}$	$\overline{\overline{A}.B + A.\overline{B}}$	Result
0	0	1	1	0	0	0	1	A = B
0	1	1	0	1	0	1	0	A < B
1	0	0	1	0	1	1	0	A > B
1	1	0	0	0	0	0	1	A = B

9.6   0101

10 Microprocessors and computing systems

Summary

Number systems, binary, denary, octal, hexadecimal. Binary arithmetic. Microprocessor instructions. Operator and operand. Minimum computer system. ROM, RAM and I/O. Bus lines. Interrupts. Microprogram and instruction set. Registers. Volatility of memory, backing stores. Floppy and hard disks. Bus drivers and transceivers. Modems. Assembler language. Servicing problems.

Note: Throughout this chapter, the letter K means 1024, and M means $1024 \times 1024 = 1048576$. These abbreviations are used in computing in place of the more usual $k = 1000$ and $M = 1\,000\,000$ because 1024 is 2^{10} and 1048576 is 2^{20}.

Practical note: This section requires experience with a simple 8-bit microcomputer system with a hex keypad and LED/LCD display. In addition, some units should be available on which faults can be placed in a way that is not obvious to the user. The recommended faults are o/c to supply positive, RESET pin connected to level O, o/c clock input to CPU, read/write line connected to + 5V, o/c or s/c address or data lines

A microprocessor is a programmable logic chip which can make use of memory, meaning chips that contain units like flip-flops that will remain in one selected state, 0 or 1, until altered. The microprocessor can address memory, meaning that it can select stored data from a particular part of memory and make use of it, or place such data into memory at an address chosen within the microprocessor. Within the microprocessor chip itself, logic actions such as the standard

278

NOT, AND, OR and XOR actions can be carried out, as well as a range of other actions such as shift and rotate (the straight-ring counter action), and some simple arithmetic. The fact that any sequence of such actions can be carried out under the control of a program is the final item that completes the definition of a microprocessor.

In general, microprocessors are designed so as to fall into one of two classes. One type is intended for industrial control, and this also extends to the control of domestic equipment, such as central heating systems. A microprocessor of this type will often be almost completely self-contained, with its own memory built in, and very often need to work with a limited number of binary digits (called bits) at a time, perhaps 4. The number of possible programming instructions need only be small. The control microprocessor will also be offered typically as a 'semi-custom' device, with the programming instructions put in at the time of manufacture for one particular customer. By contrast, the alternative is the type of microprocessor whose main purpose is computing. The computer type contains little or no memory of its own, but is capable of addressing large amounts of external memory. It will deal with at least 8 bits, and more usually 16 or 32 bits of data at a time. It has a much larger range of instructions, and will generally operate at high speeds, 16MHz or more.

Numbers and number systems

A microprocessor system operates on binary numbers, using some as instructions and others as data, but the simple binary number system as has been described in Chapter 9 is seldom the most convenient way to express a number, mainly because the base is a small number, 2.

The major problem of using a small number as a base is that any reasonably large number requires a long string of symbols to represent it. With a binary system that is restricted to 8 bits in length the maximum quantity that can be represented is $2^8 = 256$. By extending the binary system to include two 8-bit groups (bytes) per number, it becomes possible to use *standard form* representation for very small or very large numbers. For example, $300\,000\,000 = 3 \times 10^8$, where 10^8 is described as the *exponent* part of the number.

Binary coded decimal

Several coding methods are available that can be used to overcome an awkward situation, such as when numbers need to be continually displayed in their decimal form, whilst being processed in binary. The one commonly used is known as the *binary coded decimal* (BCD) system, where each denary digit of a number is represented by its 4-bit binary equivalent. The BCD/Decimal equivalents are listed in Table 10.1. Thus the decimal number 892 would be repre-

Table 10.1

Decimal	Binary coded decimal
0	0000
1	0001
2	0010
3	0011
4	0100
5	0101
6	0110
7	0111
8	1000
9	1001

sented by 1000,1001,0010. In practice, the interchange between the two forms could be performed by encoder/decoder circuits similar to those described earlier.

Signed numbers

Just as the denary system can be used to represent positive, negative or fractional numbers, so can the binary system. In this case it is common to use the leading bit (first or left-most in the string) to represent the polarity (positive or negative) of the number. By convention, a leading zero signifies a positive number. Thus, the 4-bit word, 0101 is $+5$, whilst 1101 is -5. Computers commonly work with 8-bit bytes, so that using this concept, operation is restricted to the range $-2^7 = -128$ to $+2^7 = +128$. However, this range can be considerably extended using the exponent concept.

Other coding methods

It has already been pointed out that using a binary code requires a large number of bits. If a human operator has to enter a large amount of data into a computer system in this form, it will take a long time. In addition, the tedium will lead to operator errors. What is needed is a short code for the operator to use and one which the computer can easily convert into binary itself.

Octal coding

If a string of binary bits is divided into groups of three, each group represents one of $2^3 = 8$ different numbers. These are shown in Table 10.2 and are very

Table 10.2

Binary	Octal
000	0
001	1
010	2
011	3
100	4
101	5
110	6
111	7

similar to the first eight entries in the BCD table.

From this table, the binary string of number 111,010,101 can be represented 725. To distinguish such an octal number from a decimal one it is common to add a suffix 8, that is, 725_8. With this coding, the number of bits for data entry is reduced and the computer can easily be programmed to convert it into binary.

Exercise 10.1

1   Convert the decimal number 28 into both binary and octal:

$$(011\ 100,\ 34_8)$$

2   Convert the octal number 762 into both binary and decimal:

$$(111\ 110\ 010, 498_{10})$$

Hexadecimal coding

The fact that an 8-bit byte does not divide exactly into groups of 3 bits creates problems for the computer. This can be overcome by using a code based on $2^4 = 16$ (hexadecimal). Now an 8-bit byte divides exactly into two hexadecimal numbers. The only problem is how to represent the decimal numbers 10 to 15 with a single symbol. A standard has been agreed that uses the first capital letters of the alphabet as shown in Table 10.3.

To convert from binary into hexadecimal coding, each byte is divided into two groups of 4 bits and each group then replaced by the character shown in Table 10.3. Thus 1101,0100 becomes D4.

The weighting of each character and its hexadecimal equivalent are shown for up to two character hexadecimal numbers in Table 10.4.

Table 10.3

Decimal	Binary	Hexadecimal
0	0000	0
1	0001	1
2	0010	2
3	0011	3
4	0100	4
5	0101	5
6	0110	6
7	0111	7
8	1000	8
9	1001	9
10	1010	A
11	1011	B
12	1100	C
13	1101	D
14	1110	E
15	1111	F

Thus $2E = 2 \times 16^1 + E \times 16^0$
$= 32 + 14$
$= 46_{10}$ and
$A4 = A \times 16^1 + 4 \times 16^0$
$= 10 \times 16 + 4 \times 1$
$= 160 + 4$
$= 164_{10}$

Exercise 10.2

1 Convert the hexadecimal number A8 into binary and decimal numbers:

$$(1010,1000.168_{10})$$

2 Convert the decimal number 216 into binary and hexadecimal numbers:

$$(1101,1000.D8)$$

Addition

The rules for octal and hexadecimal addition are best stated in tabular form. Table 10.5 shows that for hexadecimal addition, where the sum obtained by

Table 10.4

Hexadecimal digit	Exponent value	
	1	0
0	0	0
1	16	1
2	32	2
3	48	3
4	64	4
5	80	5
6	96	6
7	112	7
8	128	8
9	144	9
A	160	10
B	176	11
C	192	12
D	208	13
E	224	14
F	240	15

adding the head of any row to the head of any column is given at the intersection. Make up a similar table for octal numbers, remembering that the largest digit in octal is 7.

Subtraction

The rules for binary subtraction can be stated as follows:

$$0 - 0 = 0$$
$$0 - 1 = -1$$
$$1 - 0 = 1$$
$$1 - 1 = 0$$

The negative sign of the second entry creates a problem for a computer system so that this operation is better carried by the method of *complements*. If a number x is to be subtracted from another number y, x is first converted into an equivalent negative number and then added to y to give the required result.

This method can be demonstrated using a decimal number example.

By the normal method of subtraction, $96 - 72 = 24$. The complement of 72 is

Table 10.5

0	1	2	3	4	5	6	7	8	9	A	B	C	D	E	F
1	2	3	4	5	6	7	8	9	A	B	C	D	E	F	10
2	3	4	5	6	7	8	9	A	B	C	D	E	F	10	11
3	4	5	6	7	8	9	A	B	C	D	E	F	10	11	12
4	5	6	7	8	9	A	B	C	D	E	F	10	11	12	13
5	6	7	8	9	A	B	C	D	E	F	10	11	12	13	14
6	7	8	9	A	B	C	D	E	F	10	11	12	13	14	15
7	8	9	A	B	C	D	E	F	10	11	12	13	14	15	16
8	9	A	B	C	D	E	D	10	11	12	13	14	15	16	17
9	A	B	C	D	E	F	10	11	12	13	14	15	16	17	18
A	B	C	D	E	F	10	11	12	13	14	15	16	17	18	19
B	C	D	E	F	10	11	12	13	14	15	16	17	18	19	1A
C	D	E	F	10	11	12	13	14	15	16	17	18	19	1A	1B
D	E	F	10	11	12	13	14	15	16	17	18	19	1A	1B	1C
E	F	10	11	12	13	14	15	16	17	18	19	1A	1B	1C	1D
F	10	11	12	13	14	15	16	17	18	19	1A	1B	1C	1D	1E

obtained by subtracting each digit in turn from 9 to obtain 27. Adding 1 gives the complement of 72 as 28. Now $96 + 28 = 124$. Assuming that the adder system only works to two digits, the leading 1 will 'overflow', so that the complements method yields the same answer.

Method of two's complement for binary

Take the binary number to be subtracted say, 01011011, and invert each bit in turn to give 10100100. Now add 1, so that the two's complement is 10100101.

To evaluate $(01110000 - 00100101)$, which is equivalent to $(112 - 37)$ in decimal, first find the complement of 00100101, which is $11011010 + 1 = 11011011$. Adding now yields:

```
  01110000
  11011011
 101001011
```

working to 8-bit bytes, the leading 1 overflows so that the answer is 01001011 (75 in decimal).

Method of hexadecimal complements

To evaluate A3 − 8B, we first find the complement of 8B:

8   B (what must be added to each digit to make 15_{10})
7   4 (Add 1)
7   5 (Complement of 8 B)

A   3 +
7   5
───────
118 (leading bit overflows)

Therefore A3 − 8B = 18

Exercise 10.3

Evaluate the following using the complements method:

1   (01100000 − 00110000)
2   (C3 − 4D)

(*Answers at end of chapter*)

Binary multiplication and division

These operations are most easily carried out in *shift registers*. If a bit is moved one place to the left, its value is doubled; whilst if it is moved one place to the right its value is halved. Successive shifts therefore represent multiplication or division by powers of 2. Bits that are moved to the right of the 2^0 cell represent 1/2, 1/4, 1/8, and so on, in decimal terms.

Note that though the binary code system is the basis of representing numbers, it does not follow that the binary figures that are stored can easily be converted to denary numbers, especially when the numbers are stored in *floating-point* form, using one set of bits as a mantissa and another set as an exponent. Letters of the alphabet, digits 0 to 9 and various punctuation marks can be represented as numbers using a code called ASCII (the first letters of American Standard Code for Information Interchange), in which each character makes use of a number code in the range 32 – 127 denary.

Essential terms

Microprocessors date from about 1974, and because the microprocessor was a new product, it has developed a whole language of new words. Some of these will be explained later in this chapter, but a few need to be understood and defined at this stage so that you can follow what is being described without having to look up the meanings of words.

The idea of a bit as meaning a binary digit (0 or 1) has already been explained. Microprocessors deal with collections of bits in parallel, and one very common grouping is eight bits, termed a byte. A *nibble* is, reasonably enough, half of a byte so that if we have a byte 10011100 then its upper nibble is 1001 and its lower nibble is 1100. Memory chips are still constructed so as to cater for bytes, but most microprocessors at the time of writing are designed to use larger units, the word (16 bits, equal to 2 bytes) or the Doubleword (Dword) of 32 bits, 4 bytes.

An *instruction*, as far as the microprocessor is concerned, is a byte or set of bytes that will cause an action to be completely carried out. Instructions are coded as numbers, and consist of two parts. The first part of any instruction is called the *operator* or *op-code* and is the code that determines the action, such as add, shift or store. The second part of any instruction is the *operand*, showing what the action must work on. This could be the data, meaning the number that is to be used, but is just as likely to be an *address*, meaning a number that indicates the position of the data in the memory or wherever else it might be found. You can think of the difference as being like leaving a note that reads 'Buy consolidated microstock' and one that reads 'Ring 2722175'. The first is a direct piece of information or data, the other is indirect – it shows how the data is to be found.

Minimum computer system

A minimum-scale computer system must contain four essentials. These are:

1   The microprocessor or central processor unit (CPU)
2   Memory of the RAM type
3   Memory of the ROM type
4   Input/output (I/O) units (interfaces).

These parts are indicated in Figure 10.1 Not shown in this diagram is the clock circuit, which can sometimes be part of the microprocessor itself. This unit will supply timing pulses which are at a high frequency and practically always crystal-controlled. Each clock pulse is used to trigger a microprocessor action, so that the computer system operates synchronously.

The clock frequency is always high, several MHz, so that all the program instructions that the microprocessor will use must be held in memory waiting to be used, along with the numbers that represent the data. If there is nothing in the memory, then the CPU cannot operate; it is as useless as a record player

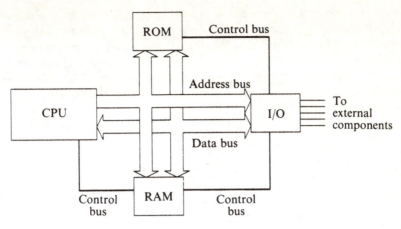

Figure 10.1   A simple CPU, ROM, RAM and I/O system

with no records. There must be instructions present in the memory permanently, ready for the CPU to use at the moment it is switched on, and there must be provision also for entering new sets of instructions so that the CPU can be made to carry out new tasks. This is why two different types of memory are used.

ROM means read-only memory, memory which cannot be altered and which the CPU can read but not change. The bits that are stored in ROM are permanent, internal connections to 0 or 1 voltage levels, so that the instant power is applied to the ROM chip, the numbers stored in the memory are accessible. Each computer system must contain some of its memory in ROM form so that the machine can be switched on and used. The ROM will contain sufficient instructions to allow other instructions to be read in, perhaps from magnetic tape but more likely nowadays from a magnetic disk, either detachable (floppy) or fixed (hard).

The RAM is memory that can be read or written, and which loses all information when the power is switched off, since it consists of units like flip-flops. The name means random access memory, dating from an earlier time when some memory systems could obtain numbers by shifting registers until the correct numbers emerged. All memory in modern systems allows random access, meaning that a byte can be obtained from any part of the memory without the need to read all of the other bytes as well. This is done by *addressing*, providing memory chips with address lines so that a number put on to the address lines in binary form will allow a specific part of the memory to be used either for reading or writing. In a ROM, of course, the data can only be read.

The I/O is the portion of the computer system that allows input or output. Input can be a set of instructions that will be placed into the memory in the form of a program that the CPU will carry out (*execute*), or it could be data obtained (very slowly) from a keyboard. Output could be information to a disk, to a VDU (display screen) or to a printer. In general, I/O is very much slower than the clock rate of the computer, so that the I/O unit incorporates some storage to allow bytes to be stored until external devices like disk drives, VDUs or printers can make use of them, or to allow bytes from external units to be stored until the CPU can attend to them. Figure 10.2 shows a complete system which incorporates these external units.

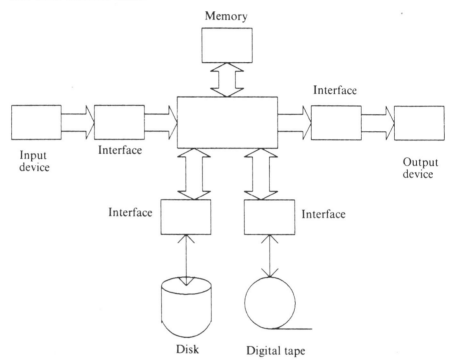

Figure 10.2 **A more elaborate diagram of a microprocessor system showing the interface units which lie between the MPU and other parts of the system**

Exercise 10.4

Using the standard 8-bit microprocessor system, identify the CPU (MPU), memory chips, display decoder, I/O port.

The buses

Because the CPU interacts with ROM, RAM and I/O, it must be connected to all of these units, and the method of connection is parallel connection, making use of a *bus*. A bus is a set of connections which are shared by a number of devices (the Latin word *omnibus* means for all, and is also the root of the more familiar no. 11 bus). Figure 10.3(a) shows a simplified 4-line bus which connects

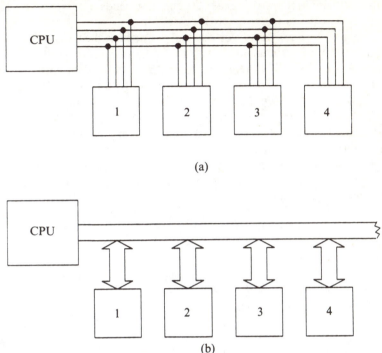

(a)

(b)

Figure 10.3   A system in which several lines are all connected to a set of units (a) can be shown in bus form (b). This latter form is the one that is used for microprocessor circuit diagrams

four units, but when a bus is shown on a diagram the individual connections are not usually illustrated (10.3(b)) unless one unit uses only a selected number of the connections.

A computer system will use at least three buses, the *address bus*, the *data bus* and the *control bus*, and of these, the address bus often consists of the largest number of lines. The older type of microprocessor, still used extensively for process control, had a 16-line address bus. This allows the lines of the bus to be used for 16-bit numbers, so that a total of 65 536 addresses could be used. If each address corresponded to a byte of data, then the data bus could be an 8-bit bus, allowing a single byte of data to be read to or written from the CPU in one instruction.

There is an important difference between these buses. The address bus is driven exclusively by the CPU, so that the CPU can control the memory address (or the I/O device) that it needs to use either for reading or writing. Such a bus is unidirectional, meaning that the signals on the bus are always in one direction, sent out from the CPU. Even if more than one CPU uses the bus, or if some other type of chip (such as a direct memory access chip) can use the address bus, the bus is still unidirectional because the chip that sends out the address signals never receives address signals from another chip.

By contrast, the data bus is bidirectional. Sometimes the CPU will read from memory or I/O, with signals travelling to the CPU from these other chips; at other times the CPU will write, providing outputs on the data lines that will be copied to memory or I/O. One of the limitations of a bus structure is that data cannot flow in both directions at the same time because of the use of the same lines; similarly only one device can be addressed at a time.

The address bus and data bus structure of different microprocessors is often quite similar, but the control bus is a collection of signal lines that will almost certainly be specific to each microprocessor type. Most control lines are unidirectional, such as the R/W line which carries signals from the CPU to memory so as to determine whether the memory is to be read or written. Other lines are used to carry signals to the CPU, and of these the 'interrupt line' is the most important. The interrupt line allows a signal to interrupt the action of the CPU so that it will complete the instruction that it is working on and then switch to a routine, called the *interrupt service routine* which will attend to the problem that caused the interrupt. One very common use of this system is for servicing a keyboard, so that pressing a key generates an interrupt, forcing the CPU to read the keyboard and store the code for the key that has been pressed. Interrupts are of two types, the *non-maskable* type which are used for emergencies only (like power failure or memory failure) and the *maskable* type such as interrupts to deal with keyboard requests. The maskable type of interrupt, as its name suggests, can be prevented from affecting the CPU, whereas the non-maskable type cannot and will override any other signals. Figure 10.4 shows a selection of control bus signals as used in some microprocessors in current use.

The data sheets for a microprocessor will state the active condition for each pin. Active high means that the action will be carried out when the pin voltage is at logic 1; active low means that the action will be carried out when the pin voltage is at Logic 0. In general, address pins are active high and control lines contain some pins which are active high and others which are active low.

Exercise 10.5

Using the standard 8-bit microprocessor system, identify the following:

(a) address bus lines
(b) data bus lines.

Find and list the lines of the control bus.

In the following descriptions of typical control signals, the signal will be either active HIGH or active LOW, and only the word *active* has been used here to avoid confusion.

RD   An output which goes active to select reading of memory or I/O
WR   An output which goes active to select writing of memory or I/O
M/IO   An output which in one state selects memory and in the other state selects I/O
WAIT   An input that forces the CPU to suspend processing
INT   An input that causes a maskable interrupt
NMI   An input that causes a non-maskable interrupt
RESET   An input that resets all registers and re-starts processing

Figure 10.4   Some of the control signals that are commonly used on microprocessors

Internal structure

A simplified version of the internal structure or architecture of a typical micro-processor chip is illustrated in Figure 10.5. This consists of the four sections

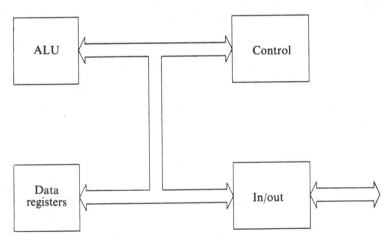

Figure 10.5   A simplified version of the internal structure of a microprocessor

labelled as ALU, control, data registers and in/out, all of which are indicated as interconnected. The control section is used to decode the program instructions (one or more bytes) and ensure that each part of a *command* is carried out. This is done by opening and closing gates at each clock pulse so as to connect different parts of the circuits together. The signals that are used for this purpose

291

are permanently stored within the chip, and are sometimes referred to as the *microprogram*. These signals cannot be altered, and their effects are described in the *instruction set* for the CPU.

The ALU is the arithmetic and logic unit of the CPU, the part where most of the work of calculations is carried out. The ALU normally provides for addition and subtraction of bytes, words or Dwords, depending on the size of the CPU, and if multiplication and division are to be carried out they will be done using shifting and adding methods under the control of a microprogram. A surprising amount of computing work does not involve the ALU, because it consists simply of copying a byte from one place to another. Word processing, for example, consists of copying bytes from a keyboard into memory, with another copy to the screen display, with no need for arithmetic unless a running count of characters is maintained.

The data registers, usually abbreviated to registers, are temporary memory units which store bytes that the CPU is currently working on. Each register will be one or more bytes in size, depending on the unit of data that the CPU uses, and the contents of a register can be shifted left or right, rotated, copied to memory, added to or subtracted from the contents of another register (using the ALU). The advantage of using registers within the CPU is that access to data in these registers is very rapid, much more rapid than reading data from memory. Virtually all instructions to the CPU involve the use of one or more registers.

Some registers are specialized. The program counter (PC) or instruction pointer (IP) register is used to store an address number. This is the address in memory of the next instruction that the CPU will work on, and this address number is adjusted each time a new instruction is read in. The flags register is another specialized store which keeps a collection of bits that signal the effect of the most recent operation. One bit, for example, will be set if the most recent action made the contents of a register zero; another will be set if the contents of a register became negative (most significant bit set). Figure 10.6 shows the block diagram in more detail, illustrating the registers.

Finally, the in/out part of the CPU deals with the bus signals, sending out the address number on the address bus lines, reading or writing data on the data lines, and sending or receiving control signals along the lines of the control bus. A few of these control signals have already been mentioned and in addition there will be a RESET input line. The effect of activating this line is to restart the CPU, clearing all of its registers, resetting the memory address number to its starting value, and finally forcing the CPU to carry out a read action for an instruction. This enables the computer system to be started when the machine is switched on, and also to re-start if a badly written program has caused problems. The designer of the system will have to ensure that the address which is used immediately following a restart is that of a suitable set of instructions contained in a ROM chip. These instructions must be in ROM so that they are always present and cannot be altered.

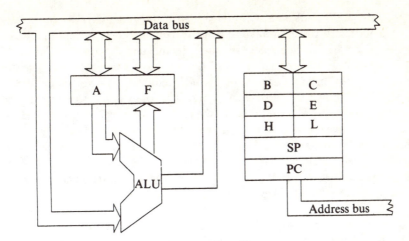

Figure 10.6 **A diagram of the internal structure of the 8080 type of 8-bit microprocessor, showing the registers, ALU and internal buses**

The fetch–execute cycle

The fetch–execute cycle of the CPU is the sequence of actions that are needed to carry out an instruction. This should not be confused with the clock cycle. The clock cycle is the time needed for one complete clock pulse, and is the timing unit for the CPU. All CPU actions will require several clock pulses, because each part of an action will be synchronized to a separate clock pulse. Some actions may be completed in 2–4 clock pulses; others may take 40–80 clock pulses to complete, depending on how complicated the action is.

A typical fetch–execute cycle starts with the address of an instruction being placed on to the address lines during one clock cycle. The source of the address number is the IP or PC register, and the number will often be obtained simply by adding 1, 2 or 3 to the address number that was previously stored in the register. At the same time, the R/W control signals will be set for reading memory. The address number may have to be held, by a set of latches if need be, for some time so that the memory can respond, allowing for the access time of the memory. At a later clock pulse, then, the instruction code will be available on the data bus, copied from the memory chips, and this can be read into a *program register*, which is part of the control section of the CPU.

The byte or word in this register is then analysed and used to connect up the appropriate microprogram. A few instructions require no data to work on, and so would be executed at this point, but for most instructions, the first byte or word is the *operator* and needs to be followed by one or more *operands*, data that will be used in the instruction. These data might be a number to be used directly, a code for a register, or an address in the memory. In following clock cycles, these additional data will be read, with the address lines providing the memory address if the data have to be read from memory. If data in a register are used, the code for the register is usually included in the operator, so that the extra step is not required. If a byte or word has to be read from memory, another read action must now be performed, requiring more clock cycles.

The data are now used. If the instruction was, for example, to ADD, the data will be sent to the ALU to be added to the contents of a register, and the register code would have been included in the instruction code. The result of the addition will be placed, usually one clock pulse later, into the same register, completing the execution part of the cycle. The CPU will by that time have altered the number in the IP or PC register ready to read the next instruction.

The points to note here are:

1 Several steps are involved in even the simplest CPU action
2 Each step requires one or more clock cycles
3 Many actions require memory to be read more than once
4 The action of the IP/PC is automatic, so that the bytes must be placed into memory in the order that the CPU requires.

Exercise 10.6

Use the standard 8-bit microprocessor unit. Enter a simple test program into memory and single-step through the program. For each step, note the contents of the main registers (program counter and accumulator in particular).

Storage

The CPU is totally useless without a program to read and execute, so that the performance of any system depends completely on having a program stored in memory. We have seen already that RAM memory loses all of its information when power is switched off; this type of memory is called *volatile*. Some method must therefore be used to retain program data to load into the RAM so that programs can be used, and such devices are termed *memory backing stores*. Typical methods are magnetic disks or tapes. Their action is much slower than that of memory within the computer, but they retain their stored information when the machine has been switched off.

The memory chips within the computer or other microprocessor system will be ROM, RAM or some version of PROM, and before we look at these in detail we need to know what characteristics of memory are important. All memory chips will have address pins so that one unit in the memory can be selected, and also one or more data pins. In addition, the chips will have some method that activates the chip (a chip-select pin) and in the case of RAM, a pin that allows the action to be switched between reading and writing. The important features are:

- volatility
- access time
- data transfer rate
- static or dynamic operation

and of these we have already mentioned volatility. A volatile memory, like RAM, will lose its stored data when power is removed from the chip. Non-volatile memory, like ROM or PROM, does not lose data, and this is true also of the backing store, disk or tape.

The access time of a memory chip is the time that elapses between establishing the address bits and obtaining access so that data can be read or written. A typical access time is 100ns, and though this might seem short, remember that a clock rate of 16MHz, quite common nowadays, allows only 62ns between clock pulses; a few desktop computers use clock speed of 33 MHz, and larger machines can use even higher speeds.

Fast-access memory is very expensive, and manufacturers have devised ways of allowing comparatively slow memory to be used with fast clock rates. One method is adding 'wait states', meaning that on each address cycle, the address will be held for one or more clock cycles longer than normal to give the memory time to respond. Another method is the use of 'cache memory', a small piece of very fast memory which is used by the CPU with data from the main memory being copied to the cache at intervals. This works because most programs operate by repeating a set of instructions, and if all of those instructions are held in the cache memory, they can be executed at full speed. Yet another method is pre-fetching or *pipelining*, in which the CPU fetches data for an instruction while the previous instruction is being completed. This can be combined with memory-banking, using different parts of memory for the different parts of an instruction so that each bank of memory is accessed at less frequent intervals.

The data transfer rate is the rate at which data can be read or written once access has been made. Once again, this requires either the use of very fast-acting memory or the methods just described to allow the use of fast clock rates.

Memory chips

ROM consists of permanent connections made inside the chip, with each connection (to 1 or 0) made to the appropriate data pin when an address is selected on the address lines. ROM is non-volatile, and ROM chips are usually arranged in byte units, such as 16K × 1 byte (a total of 131 072 stored bits) which can be connected directly to the address and data bus of the system. The access time of ROM is usually significantly slower than that of RAM (though data transfer rate is as fast), and in some computers the contents of the ROM are copied into RAM ('shadow-ROM') just after the machine is switched on, so that routines in the ROM can be used at a higher speed.

An intermediate stage between ROM and RAM is PROM, programmable read-only memory, which exists in several different varieties. A PROM is a ROM that can be written and which is non-volatile. A typical early type of PROM was the fusible link type, in which each address gave access to a point that was connected both to V + through a miniature fuse and to V − through a resistor. By applying a higher than normal voltage, the fuse at each addressable point could be blown so that each addressable point would from then on be at Logic 0 or the fuse could be left as it was, giving Logic 1, since the resistance of the fuse was much lower than that of the resistor. Since normal + 5V supplies were incapable of blowing links, these PROMS were read-only, but once programmed they could not be reprogrammed other than by severing other links to make more zero bits. PROMS have long access times, like ROM and data transfer rates that match those of RAM chips.

Later PROMS are of the EPROM variety, with the E meaning erasable. EPROMS are constructed by using higher than normal voltages to create conducting paths in lightly-doped silicon, establishing a 0 or 1 at each addressable point. Once again, because normal operating voltages have no effect, these paths are permanent so that the chip is read-only. All the paths, however, can be erased by exposing the silicon to ultra-violet light through a quartz window, using a short wavelength UV which is obtained from a special fluorescent tube. This UV is very dangerous to the eyes, and the PROM erasing units called *prom washers* cannot be switched on until the lid has been closed to prevent any leakage of UV.

Another form of PROM is the EAROM, electrically alterable ROM, in which the data are retained even with power off, but which can be written like an ordinary RAM chip when a write-select pin is activated. The access time is much the same as that of ROM.

RAM memory

RAM memory is volatile, losing stored information when power is switched off. There are two main types of RAM, however, static and dynamic. Static RAM uses units that are flip-flops, so that current has to be drawn by each unit whether it is storing a 1 or a 0. Dynamic RAM uses miniature FET capacitors, with a charge stored in a capacitor indicating 1 and zero charge indicating 0. The problem is that the very small capacitors that have to be used do not retain charge, so that each capacitor that stores a 1 in a dynamic RAM needs to have its charge *refreshed* at intervals of about one millisecond. This refreshing is carried out automatically as the addresses on the chip are selected, so that a set of dynamic memory chips must have their addresses selected at intervals of not less than 1ms whether they are being used for reading and writing or not. This action is carried out by dynamic RAM controller chips.

The additional complications of using dynamic RAM are worth while because of the very large memory sizes that can be constructed and the very low power needed for operation. Current is required only for writing and refreshing, and can be very small, a few milliamps for a chip that might store typically 64K, 256K or 1024K of bits. By contrast, static RAM requires much higher currents, consequently dissipating more heat and imposing a limit on the number of bits that can be stored. Static RAM is also considerably more expensive to manufacture. Small size static RAM sets are used for cache memory (see earlier) and also as CMOS RAM for battery-backed storage. This latter use allows essential items of data to be held even when the main power is switched off, by using a backup battery supply from which the current drain is very small. The battery can also be used to power a clock so that time is correctly maintained even with the computer switched off.

Exercise 10.7

Use the standard 8-bit microprocessor system, with a RAM-test program provided or ready to be entered. Single-step through this program, noting the contents of registers. Use the monitor system to find the hex code of data stored at the first of the tested RAM addresses. Increment the program counter and find the data in the second RAM location. Repeat until the contents of the first ten RAM addresses are noted.

Backing storage

Backing stores use magnetic media, almost always in the form of magnetic iron oxide deposited in disks or tapes. Desktop computers generally use disks, but tapes are still used for large (mainframe) computers and as a way of secondary back-up for disks. The main disadvantage of tape is that it is necessary to wind a reel of tape through a reader to find the start of a set of stored bytes. This type of storage is called *sequential*, because all the bytes of a set are stored on the tape in sequence, and must be read in sequence. On a disk, by contrast, reading can be at random, starting at any point on the disk. All these forms of backing store are non-volatile, however.

Before we can appreciate the differences between the floppy type of disk and the hard disk, we need to be as clear about the floppy disk unit itself and how it works, as about the hard disk unit. When you insert a floppy disk into a drive, and activate the drive, a hub engages the central hole of the disk, clamps it, and starts to spin it at a speed of about 300 revolutions per minute. The disk itself, Figure 10.7, is a circular flat piece of plastic which has been coated with

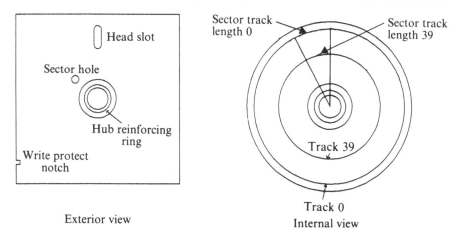

Exterior view | Internal view

Figure 10.7 **A typical floppy disk, showing external view of the cardboard cover, and the internal view of the disk itself with its organisation into tracks and sectors**

magnetic material on each side. It is enclosed in a floppy plastic envelope to reduce the chances of damage to the surface, and the hub portion of the disk is also built up in plastic to avoid damage to the disk surface when it is gripped by the drive. The surface of each disk is smooth and flat, and any physical damage, such as a fingerprint or a scratch, can cause loss of recorded data. The jacket that surrounds the disk has slots cut into it so that the disk drive head can touch the disk at the correct places and also a hole for locating the disk position.

Through the slot that is cut in the envelope, the heads of the disk drive can touch the surface of the disk, one head on each side. These heads are tiny electromagnets, and each head is used both for writing and for reading data. When a head writes data, electrical signals through the coils of wire in the head cause changes of magnetism. These in turn magnetize the disk surface. When the head is used for reading, the changing magnetism of the disk as it turns causes electrical signals to be generated in the coils of wire. This recording and replaying action is very similar to that of a cassette recorder, with one important difference. Cassette recorders were never designed to record digital signals from computers, but the disk head is. The reliability of recording on a disk is therefore very much better than could ever be obtained from a cassette, which is why computers for serious use never feature the use of ordinary audio cassettes. The same system of recording and replaying heads is used for hard disks as for the floppy type, but the larger-capacity hard disk units use multiple disks with two heads per disk.

298

Unlike the head of a cassette recorder, which does not move once it is in contact with the tape, the heads of the disk drive move over short distances, driven by stepper motors. If the head is held steady, the spinning disk will allow a circular strip (sometimes referred to as a 'cylinder') of the magnetic material to be affected by the head. By moving the head in and out, to and from the centre of the disk, the drive can make contact with different circular strips of the disk. These strips are called 'tracks' and unlike the groove of a conventional record, these are circular, not spiral, and they are not grooves cut into the disk. The track is invisible, just as the recording on a tape is invisible. What creates the tracks is the movement of the recording/replay head of the disk drive. A rather similar situation is the choice of twin-track or four-track on cassette tapes. The same tape can be recorded with two or four tracks depending on the heads that are used by the cassette recorder. There is nothing on the tape which guides the heads, or which indicates how many tracks exist. The number of tracks that are used therefore depends on the design of the disk drives, and the standard for the older PC (IBM compatible) type of computer is to use 40 tracks on each side of a 5.25″ disk. More modern computers use 80 tracks on each side of either a 5.25″ or a 3.5″ disk.

Each of these tracks is also invisibly divided up. The reason for this is organization – the data are divided into 'blocks', or sectors, each of 512 bytes. Each track of the disk is divided into a number of 'sectors', typically 9, and each of these sectors can store 512 bytes of data. The sectors are 'marked out' magnetically, a system called 'soft-sectoring'. Each 5.25″ disk has a small hole punched into it at a distance of about 25mm from the centre. There is a hole cut also through the disk jacket, so that when the disk is rotated, it is possible to see right through the hole when it comes round. When the disk is held in the disk drive, and spun, this position can be detected, using a beam of light. This is the 'marker', and the head can use this as a starting point, putting a signal on to the disk at this position, and at eight others, equally spaced, so as to form sectors. This sector marking has to be carried out on each track of the disk, which is part of the operation that is called 'formatting'. The 3.5″ type of disk uses a different method, using a slotted key so that the position of the disk relative to its driving shaft is always fixed.

The division of the floppy disk into tracks and sectors is invisible to the user in the sense that storage is always allocated by the computer, making use of its disk operating system (DOS) program. The user does not, for example, have to specify where data are placed on a disk, though programs to do this can be written, and utility programs can be obtained which will show where each byte of data is located on a disk. Typical disk storage space sizes are 360K for an older type 5.25″ disk and 720K for the 3.5″ disk. More recently, high-density disk storage systems use 1.2M on 5.25″ disks and 1.4M on 3.5″ disks, but these storage capacities demand a high standard of quality of magnetic coating, making blank disks for these drives considerably more expensive.

Hard disks

The floppy type of disk has several limitations. The main limitation is that the disk spends most of its life out of the drive, subject to the dust and smoke in the room where the disk is housed. This, along with the quality of the disk surface, means that the recording and replaying heads of the computer, though touching the disk surface, do not necessarily make good or consistent contact with the surface. At the same time, the disk cannot be spun at a very high speed because of the friction of the sleeve which partially protects it, and the friction between the heads and the disk surface. The distance between the head and the disk restricts the speed at which data can be written and recovered, and the comparatively slow speed of the disk has the same effect.

The hard disk is a way of obtaining a much larger amount of information packed into the normal size of a disk. The disk itself and its magnetic read/write heads are sealed inside a container that ensures a dust-free environment for the disk. This sealing into a clean dust-free space allows the gap between the head and the disk to be made much smaller than could be contemplated otherwise – the head actually floats on a thin film of air between the head and the disk. This very small gap between the head and the disk allows a high packing of information on to the disk without friction between the heads and the fast-spinning disk, so that the most obvious effect of using a hard disk is the much greater number of bytes that can be stored.

Unlike the comparatively slow-spinning floppy disk, which cannot be rotated fast because the disk rubs against its envelope, the hard disk spins at a very high speed, around 3600 revolutions per minute. This means that the rate at which data can be written to the disk or read from it is much greater, some twelve times as fast as a floppy disk. Modern hard disks are mainly 3.5″ in diameter, permitting lower costs, and faster access to data, because the head needs to move over a much shorter path on a small disk.

The storage space on a hard disk is very much greater than that of a floppy, and since multiple-disk units can be used with all the disks on one spindle it is easy to manufacture units with capacities of 32M and more – 200–900M hard disks are currently available. When very large amounts of data are stored in this way, it is important to maintain additional copies, called back-up copies, on other media such as tape. Videotape is particularly convenient for this purpose, given a suitable recorder adapter. The need for back-up is because of the small gap between the heads and the disk, which could allow the heads to touch the disk if the unit were knocked or vibrated badly. Such an impact would tear off the magnetic coating, possibly making the whole of the disk, or a large part of it, useless.

Another form of back-up storage that is now available is CD ROM, using the principles of digital compact disk storage to provide 650M disks which are at present read-only. Much effort is being expended on read–write versions of these disks which will then offer a considerable challenge to the larger sizes of magnetic disks for storage.

Other circuit components

The basic computer diagram of Figure 10.2 has omitted all details, and some of these details can now be better appreciated. One important point is three-state operation, which is an essential part of bus action. At some parts in a cycle, pins on the CPU are acting as outputs, but at other times they need to be isolated so that other chips can place signals on the lines. This isolated state is sometimes regarded as a third state (the other states being logic 0 and logic 1), and chips called three-state buffers (or tri-state buffers) allow a whole set of pins to be isolated in this way. Microprocessors usually incorporate three-state isolation to most of the pins on the chip, and pins will be automatically put into the isolated state at times in a cycle when they are not being used for inputs or outputs.

The clock input also has not been noted in detail. The clock circuit is an astable whose frequency is crystal-controlled. The clock circuits may be part of the microprocessor or incorporated into a separate chip, but the crystal is always external so that the user can determine what clock rate is used. Some systems use an oscillator frequency which is divided down before being used as a clock, so that the use of a 44MHz crystal, for example, does not imply a 44MHz clock rate.

The CPU itself may not have enough drive current available to cope with the load of many devices on a bus. This can be solved by using intermediate chips which can be latches, or simply unity-gain amplifiers termed *bus drivers*. For data lines, these drivers have to be arranged so as to provide buffering in each direction, and such units are called *transceivers*.

The in/out section of a computer system normally consists of several chips of quite different types, with the common name of ports. The keyboard requires a port for passing data into the computer, the VDU requires the use of another port for displaying data from the computer. A port for a floppy or hard disk will need to be bidirectional, passing data in either direction. An output-only port is needed for the printer, and for connecting to other computers, either directly or over telephone lines, a serial port, also bidirectional, is needed.

These functions could be provided by using a number of chips, buffers, latches, registers and multiplexers, but it is more common to take advantage of specialized ICs. One such specialized IC is the PIO (programmable input output) chip, also known as a PIA (programmable interface adapter). The internal layout of a simple 8-bit PIO is illustrated in Figure 10.8.

The chip provides two 8-bit ports, with a register that controls the actions of each port and a multiplexer to ensure that the ports are linked to the data bus of

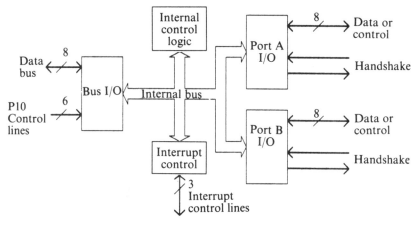

Figure 10.8   A typical PIO type of port, in this case providing two sets of ports for printers, keyboard or other uses. The Handshaking lines ensure that data is exchanged only when both systems are ready

the CPU. The control section of the chip uses the control bus of the CPU to determine when signals are passed through the port in either direction, and these control signals, along with instruction bytes which can be stored into the registers, determine how the chip will work. It is possible, for example, to make one port output only and the other input only, or even to control specific pins so that bits 0–3 are outputs and bits 4–7 are inputs in one port. The control signals from the control bus will determine when the ports are to be read or written.

PIO/PIA chips of this basic type are used mainly for parallel data interchange, in which the units are bytes or words. They are found interfacing the system buses to keyboards, VDUs, printers and disk drives, though it is more normal to find a specialized disk controller chip controlling all aspects of disk use including the port actions.

The other type of interface is the serial type in which single bits are passed in sequence along one line. This type of interface requires the use of a shift register so that a complete byte can be stored and fed out one bit at a time, or bits read one at a time and assembled into a register to be placed on the data bus of the CPU. The rate of serial bit transfer is very much lower than the clock rate of the CPU, typically 1200–19 200 bits per second.

The type of chip that is used for serial input/output is the UART, universal asynchronous receiver/transmitter, also known as ACIA (asynchronous communications interface adapter). The asynchronous part of the title does not imply that the signals are not timed, but that signals need not be sent in each time unit. In other words, if the rate is 2400 bits per second, you can send ten bits, then wait for some time and send another batch, rather than having to send 2400 bits every second. A few UART chips permit synchronous use as well and are described as USART chips. Strictly speaking, the UART is an earlier version of the ACIA chip which needed separate ports to interface with the computer, but modern units include all such porting internally so that the distinction is now unimportant.

A typical internal layout for an ACIA is shown in Figure 10.9. The multip-

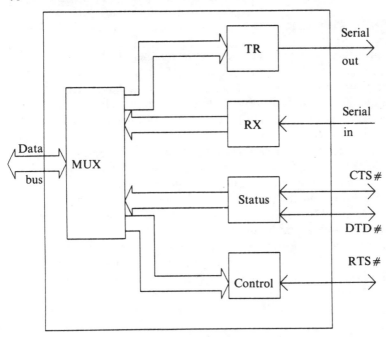

Figure 10.9   A typical serial (ACIA) type of port which converts 8-bit data to and from serial form which can be transmitted or received one bit at a time

lexer consists of registers in which bits can be read from a register or assembled into a register for serial transmission or reception. Bits to be transmitted are passed by way of a buffer to a single output line, and bits received are passed through another buffer into the multiplexer (MUX, see Figure 10.9). The status and control parts control the action, ensuring that bytes are correctly assembled and dismantled. When a byte is sent along a single serial line, each byte must be preceded by a start bit and ended with one or two stop bits, so that each byte can be separated from the next and the previous byte. The status and control units ensure that this is done correctly, and can be set to provide the correct rate of transfer, number of stop bits and other standards (called the serial *protocols*).

ACIA chips will provide signals that can be used with *modems*. A modem, an abbreviation of modulator–demodulator, is a unit which changes a Logic 0 signal into one tone and a Logic 1 signal into another tone; an alternative is to change the phase of a signal for each logic level. By using a modem, a set of logic signals can be changed into a modulated sine wave which can be sent over telephone lines or other lines which cannot accept sharply rising and falling waveforms like digital pulses.

Microprocessor types

Microprocessors are classed mainly by the number of bits that can be handled on the internal data bus. Some confusion arises here, because some microprocessors can handle only 8 bits at a time on the data bus, but 16 or even 32 bit units internally. Manufacturers tend to use the larger number where there is any discrepancy. The original microprocessors were 4-bit machines, but were almost immediately superseded for computing purposes by 8-bit machines, a group which includes many microprocessors that are still in use, including the 8080 and 8085 from Intel, the Z-80 from Zilog, the 6502 from Mostek and the 6800 from Motorola. The introduction of 16-bit microprocessors, such as Intel's 8086 and 80286 and the Motorola 68000 series set new standards, and at the time of writing 32-bit processors such as the Intel 80386 and 80486 are being used in the latest generation of computers.

The number of lines in the data bus is really of secondary importance, because memory is generally still organized in single-byte units, and much computing activity still uses byte units. For example, each character of a word is coded in 7 bits in ASCII code, and there is no particular advantage in working with four characters at a time. Much more significant is the ability to address large amounts of memory. The 8-bit microprocessors used 16 address lines to address 64K of memory. Later 16-bit processors used 20-bit address lines to work with 1M of memory, or 24 lines to work with 16M. The Intel 80386 can use 32 address lines, giving it the power to use up to 4096Mb (or 4Gb) of memory.

The speed of operation of a CPU is another important factor. Some early 8-bit types used clock rates of 1MHz or lower, and even when 16-bit processors started to be used the clock rate was generally around 4MHz. This speed has increased considerably due to the use of more efficient chip designs, and clock rates in the region of 16–33MHz are now in common use. The clock rate is not the only factor that determines the speed of a processor, however, and there is a trend to using processors which can execute only a small number of instructions, but in a very small number of clock pulses so that high computing speeds can be obtained even at low clock rates. Such processors are referred to as RISC (reduced instruction set computers).

Another point of difference is the number of registers in a processor. Early processors used only a very few registers, so that locations in memory had to be used for a lot of data manipulation. Since access to memory is always very much slower than access to registers within the CPU, the speed of a processor can be increased, within limits, by providing it with more registers as long as the CPU can organize these registers efficiently. The Z-80 led the way among 8-bit processors with eight 8-bit registers that could be used as four 16-bit registers, plus an alternative set of identical registers that could be used by switching over, and a set of four 16-bit registers, plus two special-purpose registers (for refreshing dynamic memory and storing interrupt address codes). The 16-bit processors allowed the use of 16-bit registers, and the Intel 8086 followed the earlier pattern of the 8080 and Z80 in allowing these registers to be split into separate 8-bit units. The 8086 has four general-purpose 16-bit registers (usable as 8 8-bit registers), four 16-bit pointer and index registers, and four 16-bit segment registers, plus the 16-bit instruction pointer and flag registers. The segment registers are used to specify which of 16 segments, each of 64K of memory, are being used, allowing the 16-bit IP register to locate an address in any part of a 1M memory. The pointer and index registers are used mainly in rapid and efficient movement of bytes from one part of memory to another. The main purpose of the design of registers, however, is to maintain compatibility with the older 8-bit processors such as the 8080.

The Motorola 68000 processors, though mostly classed as 16-bit, use 32-bit registers. There are eight 32-bit data registers and seven 32-bit address registers, along with stack pointer, PC and condition code (flag) registers. In addition to this user set of registers, there are five other registers which are used when more than one task is running. This large number of 32-bit registers make the Motorola processors very fast even at low clock speeds, and able to handle large amounts of memory without the need to divide the memory into segments.

The Intel 80386 uses a set of registers very similar to those of the 8086, but 32 bits wide, with the facility to use registers as 8-bit, 16-bit or 32-bit in order to allow programs written for earlier processors to run on the later types. This is a particular feature of Intel designs that has made it possible for the design of computers such as the IBM PC to become standardized, so that even very recent designs of computer can run software written ten years ago for an older design of processor. Machines that are designed around the Motorola 68000 chips include the Apple Mackintosh, the Commodore Amiga and the Atari ST, but these machines are incompatible with each other and with earlier machines.

The last point of difference between microprocessor chips is packaging. Early chips were able to use 40-pin DIL packages, Figure 10.10. The early 16-bit chips were also able to use these 40-pin DIL packages by time multiplexing the address and data lines, so that 16 pins were used for address bits on one phase and for data on the other phases of a four-phase clock. A few later designs have used 64-pin DIL packages, but a more common choice is the pin-grid array or leadless chip carrier design, illustrated in Figure 10.11. Many of these designs allow for the use of multiple supply lines, both Vcc (+5V) and Vss (earth), each of which must be connected.

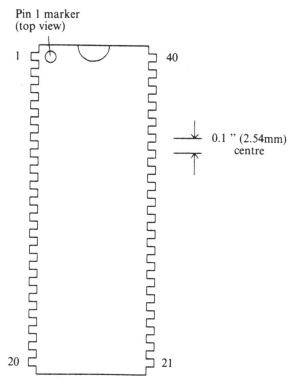

Pin 1 marker
(top view)

1

40

20

21

0.1 " (2.54mm)
centre

**Figure 10.10   The standard 40-pin DIL package which was commonly used
for the earlier types of microprocessors and support chips**

Assembler instructions

Programming a microprocessor system requires all the bytes of the program to
be placed into memory so that the system can execute the instructions whenever
the content of the IP/PC register is loaded with the address of the first byte of
the program. Programming in this way is very tedious and prone to error, even
when a hexadecimal keyboard is used to make the entry of numbers easier. For
this reason, manufacturers devise an *assembly language* for each microprocessor.

An assembly language consists of simple command words, sometimes called
mnemonics for each operator, plus agreed conventions for the way in which
numbers are used following the operator. An agreed convention is needed in
order to distinguish between a number which is to be used directly and a number
which is used as a memory reference. Numbers are almost always specified in
hexadecimal, using the suffix H as a reminder.

When a program is written in assembly language, using a word processor or
text editor, another program, the *assembler*, can convert each line of instruc-

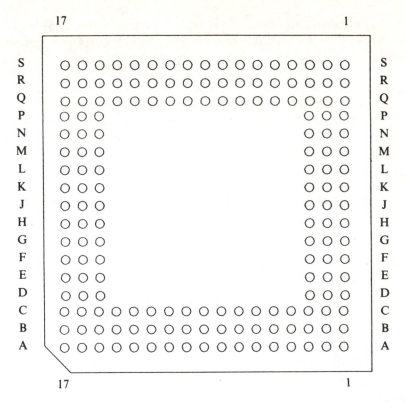

Figure 10.11   **The later style of pin-grid array format which is now extensively used for microprocessors and associated chips. Another option, not illustrated, is the leadless chip carrier, intended for surface mounting, which places all the contact points round the edge of the square package**

tions in the text into the correct codes for the microprocessor. This method of producing codes is faster and less liable to error than coding directly, because the form of the assembly language makes it easier to check what is being done. It is simpler, for example, to understand the line:

LDA, 12H (load register A with the number 12 in hexadecimal)

than the code:

3E 12

which will be placed into memory as a result of assembling the instruction line.

Three important categories of assembler instructions deal with the actions of data transfer, arithmetic and logic, and test and branch. In the following example, the assembly language of the 8-bit Z80 microprocessor has been used because the chip is still widely available and its assembly language is reasonably simple.

Data transfer on the Z80 assembly language uses the operator word LD (load) for either direction of transfer. This is similar to the use of the word MOV (move) in the later Intel 8086, 80286, 80386 and 80486 processors. The LD operator can be used for a transfer in either direction, into the microprocessor or out of it, and the difference is expressed by the order of the operands, which is destination, source. For example:

LD A, (EF00H)

means that the A register is to be loaded from memory address EF00H, whereas the form:

LD (EF00H),A

means that the memory location EF00H is to be loaded with the byte from register A. The brackets around the number indicate that this is a memory address and not a data word. Loading implies copying the byte in one location into the other location rather than moving (which would mean that the original no longer existed).

Arithmetic and logic actions mainly involve two numbers, of which one is stored in a register (using a LD instruction) and the other is specified in the arithmetic or logic instruction. For example:

ADD B, 0CH

means that the number 0CH will be added to the number previously stored in the B register. The result of this addition will be stored in the B register (or whatever destination has been specified), and if the addition causes a carry bit (a ninth bit when two bytes are added) this will set the *carry flag* in the flags register. The ADD instruction will ignore this carry bit in a subsequent addition but if the alternative instruction ADC is used then the carry will be added into the numbers. This allows numbers of more than one byte in length to be added byte-by-byte.

The arithmetic instructions of the Z80 consist of add, subtract and complement (make negative), without any provision for multiplication and division, so that these latter actions have to be carried out by program routines. The logic actions consist of AND, OR, XOR and NOT, along with register shift and rotate. The AND, OR, and XOR instructions, like arithmetic require two bytes to work on, and the logic actions are carried out *bitwise*, meaning, for example, that an AND of two bytes will be done by ANDing corresponding bits in pairs, one from each byte. The instruction:

AND C, 0FH

for example, will have the effect of masking off (removing) the upper nibble of a byte that has existed in register C, as Figure 10.12 shows.

Byte to be masked: 9CH = 10011100 in binary
To mask upper nibble use 0FH:
    10011100
    00001111   AND action
    00001100 = 0CH, upper nibble masked.
To mask lower nibble use F0H
    10011100
    11110000   AND action
    10010000 = 90H

Figure 10.12   Masking the upper nibble of a byte by using the AND 0FH instruction

The shift and rotate instructions refer to one byte only, which can be a byte contained in a register or in a memory address. The shift action can be left or right, with a 0 being shifted in at one end of the byte and a bit being lost from the other end. If, for example, we have the byte A6H, which in binary terms is 10100110 and we shift left one place, the result is 01001100, 4CH in hex. If we shift the byte A6H right one place, the result in binary is 01010011, which in hex is 53H. Note that shifting by one place is equivalent to multiplication or division by two (according to the direction of shifting) so that shifting is an important part of routine for multiplication or division.

Rotation resembles shifting, but with the bit that is 'lost' from one end of the register or memory location being put in at the other end so that the bits of the byte are never altered, only shuffled around. Rotation can be in either direction, and the Z80 instruction set allows for the carry bit to be included in a rotate action. This is convenient as a method of testing each bit in a unit by rotating each bit in turn into the carry position and testing the carry flag.

Finally, the Z80 includes many instructions of the test-and-branch type. In this context, test means that a flag bit is tested, and branch means that instead of executing the next instruction in sequence, the processor will be forced to move to another part of the program. This is a way of executing choices and of repeating actions until some condition is satisfied, like a count up or down. Some branch instructions make no tests and are described as unconditional. For example, in Z80 assembler language:

JP 23F6H

means that the program should jump to the code at location 23F6H, whereas:

JPZ 23F6H

means that the program should jump to the new address only if the zero flag is set. The zero flag might be set because a number in a register has been counted down to zero, just to take one example. All the flags in the flag register can be tested in this way and used to control jumps, but the most important tests are for zero, for negative number and for a carry.

Tests are often put in following a CMP instruction. This carries out a subtraction action but without altering anything other than the carry or zero flags. For example, CMP A,0DH will subtract the byte 0DH from the byte in register A, but the byte in A is not changed. If the bytes were equal the zero flag will be set, and if the byte in A were smaller than 0DH, the negative flag would be set.

Servicing work

In some respects, servicing microprocessor circuitry is simpler than work on analogue circuits of comparable size. All digital signals are at one of two levels, and there is no problem of identifying minor changes of waveshape which so often cause trouble in audio circuits. In addition, the specifications that have to be met by a microprocessor circuit can be expressed in less ambiguous terms than these that have to be used for analogue circuits. There is no need, for example, to have to worry about harmonic distortion or intermodulation. That said, microprocessor circuits bring their own particular headaches, the worst of which is the relative timing of voltage changes.

Digital signals are generally changing with each clock pulse, and the problem is compounded by the fast clock rates that have to be used for many types of microprocessors. An example is to find a coincidence of two pulses with a 12MHz clock pulse when the coinciding pulses happen only when a particular action is taking place. This action may be completely masked by many others on the same lines, and it is in this respect that the conventional oscilloscope is least useful. Oscilloscopes as used in analogue circuits are intended to display repetitive waveforms, and are not particularly useful for displaying a waveform which once in 300 cycles shows a different pattern. The conventional oscilloscope is useful for checking pulse rise and fall times, and for a few other measurements, but for anything that involves bus actions a good storage oscilloscope is needed. In addition, some more specialized equipment will be necessary if anything other than fairly simple work is to be contemplated. Most of this work is likely to be on machine-control circuits and the larger types of computers. Small computers do not offer sufficient profit margin in repair work to justify much diagnostic equipment. After all, there's not much point in carrying out a £100 repair on a machine which is being discounted in the shops to £50! Spare parts for the smaller home computers are also a source of worry, because many are custom-made, and may not be available by the time that the machines start to fail.

Failures in microprocessor circuits are either of components, open or short circuit, or in ICs. For ordinary logic circuits, using TTL or CMOS chips, one very useful diagnostic method relies on slow clocking. The state of buses and other logic lines can be examined using LEDs, and with a one-second or slower clock rate, the sequence of signals on the lines can be examined from any starting state. This approach is not usually available for microprocessor circuits. A few microprocessors of CMOS construction, like the Intel CHMOS 80C86, can be operated at very slow clock rates, down to d.c. This is a very useful feature, though it does not necessarily help unless you can get the buses into the state at which the problem reveals itself. As in any other branch of servicing, your work is made very much easier if you have some idea of where the fault may lie. For example, if the user states that short programs run but long ones crash, this is a good pointer to something wrong in the higher order address lines, such as an open-circuit contact on an IC holder for a memory chip. Slow clocking is not necessarily helpful for such problems, because a large number of clock cycles may be needed to reach the problem address by hardware.

For many aspects of fault finding, the use of a diagnostic program is very helpful. Such programs are usually available for computer servicing, and will help to pinpoint the area of the problem. A test program can find problems in the CPU, the RAM and the ROM, and can also point to faults in the ports or in decoder chips. Some machines, such as the IBM PC, run a short diagnostic program each time the machine is switched on, and more extensive programs can be obtained to order.

Diagnostic programs are not always available for machine-control systems, because so many systems are custom-designed. If it's likely that one particular system design will turn up several times, then a simple diagnostic program should be written, enlisting a software specialist if necessary. One very simple method is to use code that consists entirely of NOP (no operation) instructions, since this will allow the action of the address lines to be studied as they cycle through all the available addresses.

For small computers, the use of a good diagnostic program may be all that is normally needed to locate a fault. This is particularly true when the machine is one that has a reasonably long service record, with well documented problems and their solutions available for a large sample of machines. Servicing the BBC Micro, for example, is made much easier by the service history which has been built up by local education authorities on this excellent machine. By contrast, servicing of a comparatively new model may be virtually impossible through lack of information, and manufacturers can be quite remarkably uncooperative both in the provision of information and of spares. This is very often because they are working on the next model and have lost interest in last month's wonder. For machine-control circuits, easy availability of either data or spares cannot be relied on, and servicing may have to be undertaken with little more than a circuit diagram, and the data sheets for the chips that are used. The compensation here is that the chips are more likely to be standard types, with no custom-built specials.

The main problem in this respect is that servicing of microprocessor circuits cannot ever be a purely hardware operation. Every action of a microprocessor system is software-controlled, and in the course of fault diagnosis, a program must be running. For a computer system, this is easily arranged, and a diagnostic program can be used, but for a machine-control system this is by no means simple. A machine-control system, for example, may have to be serviced in situ, simply because it would be too difficult to provide simulated inputs and outputs. In such circumstances, dummy loads may have to be provided for some outputs to avoid unwanted mechanical actions. Against this should be laid the point that inputs and outputs are more easily detectable, and likely to be present for longer times, making diagnosis rather easier unless the system is very complicated. A logic diagram, showing the conditions that have to be fulfilled for each action, is very useful in this type of work. Once again, if the system is one that would have to be serviced frequently, then a 'dummy driver' which will provide simulated inputs, can be a very useful service tool.

An important point to note in this respect is the number of ways that the word *monitor* is used. The conventional meaning of the noun monitor is the VDU or screen display unit, and the verb to monitor means to observe and check a quantity. In software work, however, a monitor (more correctly a monitor program) is a form of diagnostic program which will show information such as the contents of CPU registers and system memory along with contents of specified disk sectors, allowing such contents to be changed, and the running of programs step by step. Such a program is very useful for tracing software faults.

Instruments

Of the more specialized instruments which are available for working with microprocessor circuits, logic probes, pulsers, monitors and logic analysers are by far the most widely used. A logic probe is a device which uses a small conducting probe to investigate the logic state of a single line. The state of the line is indicated by LEDs, which will indicate high, low, or pulsing signals on the line. The probe is of very high impedance, so that the loading on the line is negligible.

These probes are not costly, and are extremely useful for a wide range of work on faults of the simpler type. They will not, obviously, detect problems of mistiming in bus lines, but such faults are rare if a circuit has been correctly designed in the first place. Most straightforward circuit problems, which are

312

mainly chip faults or open or short circuits, can be discovered by the intelligent use of a logic probe, and since the probe is a pocket-sized instrument it is particularly useful for on-site servicing. Obviously, the probe, like the voltmeter used in an analogue circuit, has to be used along with some knowledge of the circuit. You cannot expect to gain much from simply probing each line of an unknown circuit. For a circuit about which little is known, though, some probing on the pins of the microprocessor can be very revealing. Since there are a limited number of microprocessor types, it is possible to carry around a set of pinouts for all the microprocessors that will be encountered.

Starting with the most obvious point, the probe will reveal whether a clock pulse is present or not. Quite a surprising number of defective systems go down with this simple fault, which is more common if the clock circuits are external. Other very obvious points to look for are a permanent activating voltage on a HALT line, or a permanent interrupt voltage, caused by short circuits. For an intermittently functioning or partly functioning circuit, failure to find pulsing voltages on the higher address lines or on data lines may point to microprocessor or circuit-board faults. For computers, the description of the fault condition along with knowledge of the service history may be enough to lead to a test of the line that is at fault. The considerable advantage of using logic probes is that they do not interfere with the circuit, are very unlikely to cause problems by their use, and are simple to use. Some 90 per cent of microprocessor system faults are detectable by the use of logic probes, and they should always be the first hardware diagnostic tool that is brought into action against a troublesome circuit. A variant on the logic probe is the logic clip which as the name suggests clips on to the line and avoids the need to hold the unit in place. Logic clipping can also be used on chips, using a clip unit which fastens over the pins of the chip and allows probes to be connected with no risk of shorting one pin to another.

Logic pulsers are the companion device to the logic probe. Since the whole of a microprocessor system is software-operated, some lines may never be active unless a suitable section of program happens to be running. In machine-control circuits particularly, this piece of program may not run during any test, and some way will have to be found to test the lines for correct action. A digital pulser, as the name indicates, will pulse a line briefly, almost irrespective of the loading effect of the chips attached to the line. The injected pulse can then be detected by the logic probe. This method is particularly useful in tracing the path of a pulse through several gate and flip-flop stages, but not if several conditions have to be fulfilled at any one time. The digital pulser is a more specialized device than the logic probe, and it has to be used with more care. It can, however, be very useful, particularly where a diagnostic program is not available, or for testing actions that cannot readily be simulated.

The logic comparator (or logic monitor) is an extension of the logic probe to cover more than one line. The usual logic monitor action extends to sixteen lines, with an LED indicating the state of each line. The LED is lit for logic high, and unlit for logic zero or the floating state. For a pulsing line, the brightness of each LED is proportional to the duty cycle of the pulses on the line. Most logic monitors have variable threshold voltage control, so that the voltage of transition between logic levels can be selected to eliminate possible spurious levels. Connection to the circuit is made through ribbon cable, terminating usually in a clamp which can be placed over an IC. For microprocessor circuits, the usual clamp is a 40-pin type, and the ends of the ribbon cable must be attached to the set of bus lines for that particular microprocessor. Pre-wired clamps are often available for popular microprocessor types. For microprocessor circuits, however, the use of a 40-point monitor is much more useful, since only a knowledge of the microprocessor pinout will then be needed. Since port chips are generally in a 40-pin package also, this allows tests on ports, which are often a fruitful source of microprocessor system troubles.

Exercise 10.8

Use the standard 8-bit microprocessor system, in working order, with a monitor program running. Using a logic probe:

(a) check the presence of a clock signal input
(b) check that the RESET input changes state when the RESET button is used
(c) check that the address and data lines are active (pulsing)
(d) check that the control lines to memory chips are active.

Logic analysers

The logic analyser is an instrument which is designed for much more detailed and searching tests on digital circuits generally and microprocessor circuits in particular. As we have noted, the conventional oscilloscope is of limited use in microprocessor circuits because of the constantly changing signals on the buses as the microprocessor steps through its program. Storage oscilloscopes allow relative timing of transitions to be examined for a limited number of channels, but suitable triggering is seldom available. Logic probes and monitors are useful for checking logic conditions, but are not useful if the fault is one that concerns the timing of signals on different lines. The logic analyser is intended to overcome these problems by allowing a time sample of voltages on many lines to be obtained, stored, and examined at leisure.

Most logic analysers permit two types of display. One is the 'timing diagram' display, in which the various logic levels for each line are displayed in sequence, running from left to right on the output screen of the analyser. A more graphical form of this display can be obtained by connecting a conventional oscilloscope, in which case, the pattern will resemble that which would be obtained from a 16-channel storage oscilloscope. The synchronization may be from the clock of the microprocessor system, or at independent (and higher) clock rates which are more suited to displaying how signal levels change with time. The other form of display is word display. This uses a reading of all the sampled signals at each clock edge, and displays the results as a 'word', rather than as a waveform. If the display is in binary, then the word will show directly the 0 and 1 levels on the various lines. For many purposes, display of the status word in other forms, such as hex, octal, denary or ASCII may be appropriate. This display, which gives rise to a list of words as the system operates, is often better suited for work on a system that uses buses, such as any microprocessor system. The triggering of either type of display may be at a single voltage transition, like the triggering of an oscilloscope, or it may be gated by some preset group of signals, such as an address. This allows for detecting problems that arise when one particular address is used, or one particular instruction executed.

A current tracer is a way of measuring current in a PCB track without the need to make a gap in the track and attach a meter. The tracer works either by sensing the magnetic field around the track or by measuring the voltage drop along a piece of track, and measurements are accurate only if the PCB track conforms to the standard 306 g/m^2 density specification, and with the tracer set for the correct width of track. Currents of 10mA or less can be measured, and this is a very useful way of finding cracked tracks, dry joints, shorted tracks and open-circuit at plated-through connections.

Finally, a signature checker allows for faults in a ROM to be checked. When a prototype system has been constructed and proved to function correctly it can be made to execute a sequence of instructions in a repetitive manner. By monitoring each node, data about the correct logical activity can be accumulated. By adding such data to the circuit or system diagram, a 'signature' for each node, under working conditions, is provided. In servicing, a faulty component can then be identified as a device that produces an error output from correct input signatures.

For computer ROMs, the definition is slightly different. When the ROM is perfect, adding all of the bytes will produce a number which is the 'signature' for that ROM, and repeating the addition should always produce the same signature if the ROM is perfect. Any deviation will indicate a fault which can be cured only by replacement of the ROM. The signature is usually obtained by a more devious method than has been suggested, keeping the size of the sum constant and omitting any overflow bits, but the principle is the same.

Exercise 10.9

Because of the complexity and variety of microprocessor control systems, guidance to any particular practical activity would have a very limited value.

However, the following are suppliers of systems training aids that, in the authors' experience, can provide a valuable practical support for the theory:

Feedback Instruments Ltd, Park Road, Crowborough, East Sussex, TN6 2QR.
Test and measuring instruments, transducers and control systems.
Flight Electronics Ltd, Flight House, Quayside, Southampton, SO3 1SE.
Microprocessor computer for systems simulation.
JJ Lloyd Instruments Ltd, Brook Avenue, Warsash, Southampton, SO3 6HP.
Educational instruments, systems lab with transducers.
LJ Electronics Ltd, Francis Way, Bowthorpe Industrial Estate, Norwich, NR5 9JA.
Microprocessor trainer with control simulation and fault simulation.
Electronics Examining Board, Savoy Hill House, Savoy Hill, London, WC2R 0BS.
Traffic lights and process-control simulators.

Use the standard 8-bit microprocessor system with one of the standard faults (see note at start of this chapter). Using the logic probe, test the signals at the microprocessor chip to reveal the fault – refer to your notes on Exercise 10.7 if in doubt.

Multiple-choice test questions

1   A microprocessor system is operating a parallel printer. This involves bytes being copied from:
    (a) CPU to ROM
    (b) CPU to PIO
    (c) CPU to RAM
    (d) CPU to UART.

2   A line in a microprocessor system is to be checked, but no program exists to run the microprocessor. Which combination of instruments would be most useful at this stage?
    (a) logic probe and current tracer
    (b) logic pulser and signature analyser
    (c) logic probe and oscilloscope
    (d) logic probe and logic pulser.

3   In a computer system, the third bit of each byte is found always to be 0 when a diagnostic program is run. This points to:
    (a) an s/c failure of a decoupling capacitor
    (b) an o/c decoupling capacitor
    (c) a faulty memory chip
    (d) a faulty CPU chip.

4   One of the following does not contribute to high operating speed of a microprocessor. It is:
    (a) a high clock speed
    (b) added wait states
    (c) use of many registers
    (d) pipelining of instructions.

5   A typical ROM memory could be described as:
    (a) volatile and dynamic, with fast access
    (b) non-volatile and static, with fast access
    (c) volatile and dynamic with slow access
    (d) non-volatile and static, with slow access.

6   A data transfer instruction uses a memory reference. This will involve signals on:
    (a) data bus and control bus
    (b) address bus and control bus
    (c) data bus, address bus and control bus
    (d) data bus and address bus.

Answers to Exercise 10.3

1   00110000
2   76_{16}

Appendix 1 Answers to test questions

Chapter 1
1. d   2. c   3. b   4. a   5. b   6. d

Chapter 2
1. c   2. a   3. d   4. b   5. d   6. b

Chapter 3
1. b   2. a   3. b   4. c   5. b   6. c

Chapter 4
1. b   2. c   3. d   4. b   5. c   6. d

Chapter 5
1. a   2. d   3. b   4. b   5. b   6. c

Chapter 6
1. d   2. b   3. a   4. c   5. d   6. b

Chapter 7
1. b   2. c   3. a   4. b   5. d   6. c

Chapter 8
1. c   2. b   3. d   4. a   5. b   6. b

Chapter 9
1. b   2. c   3. a   4. d   5. b   6. d

Chapter 10
1. b   2. d   3. c   4. b   5. d   6. c

Appendix 2 Examination structure

The following examination structure is provided for the award of the Joint Part II Certificate in Electronics Servicing of the City and Guilds and the Electronics Examination Board.

Core studies to be taken by all candidates.
224–2–11 Analogue Electronics Technology (MC) 224–2–12 to be taken by all candidates.
Practical assignments (in course) 224–2–13 Digital Electronics Technology (MC) 224–2–14 to be taken by all candidates.
Practical assignments (in course)
Options: candidates may enter either one or both of 224–2–15 Television and Radio Reception Technology (written) 224–2–16 Control System Technology (written).

Practical examination; candidates may enter one or both appropriate to the above options.

224–2–17 EEB Practical test in Television and Radio technology 224–2–18 EEB. Practical test in Control System technology.

The assessments need not all be entered in the same year. The candidate can build up a series of credits towards the ultimate award of the certificate over a period of time.

Assessment checklists

Section 01, power supplies

In a practical situation, each candidate is expected to:

- Observe safe working practices
- Measure d.c. output and a.c. ripple voltages for a power supply that consists of a full-wave rectifier and smoothing filter.
- Compare the effects of different types of filter circuit on the loaded output of a power supply.
- Study the effects on a regulated power supply as the a.c. input voltage and load current are varied separately.
- Measure the d.c. voltages and record and compare, the waveforms in a basic and a switched mode power supply, under conditions of varying load.
- Identify the effects of a single faulty component on the various types of power supply.

Section 02, amplifiers

In a practical situation, each candidate is expected to:

- Observe safe working practices.
- Measure and compare the gain, bandwidth, frequency response and distortion for various amplifiers under changing load, bias and feedback conditions.
- Demonstrate the effects of mismatching the load in power amplifiers.
- Demonstrate the signal processing of various configurations of operational amplifiers by measuring and comparing the input and output waveforms.
- Identify the effects of a single component failure on amplifier performance.

Section 03, oscillators and waveform generators

In a practical situation, each candidate is expected to:

- Observe safe working practices.
- Measure d.c. voltages and displayed output waveforms on various types of oscillator circuit.
- Identify the effects on the output waveform by changing the values of various components.
- Measure d.c. voltages and displayed input and output waveforms for various non-sinusoidal oscillators.

- Identify the effects caused by changing the value of various components.
- Measure d.c. voltages and displayed output waveforms for various timer and trigger circuits.
- Identify the effects caused by changing component values.

Section 04, basic digital circuits

In a practical situation, each candidate is expected to:

- Observe safe working practices. Verify the truth tables by testing the operation of JK bistable and tristate devices.
- Test the operation of a half adder or subtractor, a two-way multiplexer or demultiplexer and a two-bit binary comparator, which have been constructed from combinational logic gates.
- Test the operation of 4 bit serial input, serial/parallel output shift registers, 4-stage asynchronous up/down counters and synchronous decade counters.
- Identify the effect and locate a single component fault in each of the above circuits.

Section 05, basic microprocessor-based systems

In a practical situation, each candidate is expected to:

- Observe safe working practices. Measure voltages and display waveforms on the following devices in a microprocessor based system: CPU, RAM, ROM, Decoder, I/O device, Bus driver/Transceiver.
- Use the monitor program to display the contents of a block of RAM or ROM and the state of CPU registers for a given address.
- Enter a simple program to test RAM by loading a block with given data to verify success of the operation.
- Measure the voltages on the following CPU control lines after initial switch on and compare them with a test specification: reset, interrupt, non-maskable interrupt, read/write and clock (oscillator).
- Identify the effect of a single fault on the following devices: CPU, RAM, ROM, decoder, I/O device, bus driver/transceiver.

Appendix 3 EEB Practical tests

The test is presented in two sections. The candidate is expected to demonstrate his/her competence to perform a series of analogue and digital measurements using multimeters, double beam oscilloscopes, logic probes and signal sources. In the second part each candidate is expected to be able to find faults in both analogue and digital circuits to component level.

Measurement tests

The candidate is allowed one hour in which to carry out both static and dynamic measurements on a working circuit, using multimeters and oscilloscope. He has also to demonstrate his ability to sketch the time related waveforms found at various test points in the circuit.

Fault finding

The candidate is allowed one hour in which to locate three faults in different basic systems to component level and present a comprehensive report on the tests applied, the results found and the conclusions. The faulty systems used will be appropriate to the options of Television and Radio or Control Systems.

Appendix 4 Abridged Part 2 Syllabus (index version)

Note: for complete information, the reader is referred to the City and Guilds Part 2 Syllabus.

Part 2

The student is expected to be able to demonstrate competence in setting up, testing and diagnosing faults in various sytems. He must also demonstrate a general command of basic electrical and electronic principles.

01, Power supplies (Chapters 1, 2, 3, 5 and 6)

1.1. Describe the principles of loading, filters, regulators and voltage doublers.
1.2. Use block diagrams to explain the operation of switched mode power supplies.
1.3. Describe the principles of voltage controlled rectification.
1.4. Recognize the importance of component rating and radio interference in power supplies. Describe the operation of battery charging equipment.

08, Science background

Sections 002, 003, 004, 005, 006   Differentiating and integrating circuits, measuring instruments, transducers, electrical principles, integrated circuits, transformers and displays.

02, Amplifiers (Chapters 4 and 5)

2.1. Describe the operation and properties of single and multi-stage amplifiers.
2.2. Describe the operation and characteristics of power amplifiers.
2.3. Describe the operation, characteristics and applications of operational amplifiers.

08, Science background

Sections 002, 003, 004, 005, 006 Differentiating and integrating circuits, measuring instruments, transducers, electrical principles, integrated circuits, transformers and displays.

03, Oscillators and waveform generators (Chapter 7)

3.1. Describe the conditions required for oscillations.
3.2. Describe the operation of various types of oscillator. State the bias requirements and sketch time related waveforms.

08, Science background

Sections 002, 003, 004, 005, 006 Differentiating and integrating circuits, measuring instruments, transducers, electrical principles, integrated circuits, transformers and displays.

04, Basic digital circuits (Chapter 9)

4.1. Describe positive and negative logic.
4.2. Describe the principles of combinational logic circuits.
4.3. Describe the operation of monostable, bistable and astable devices.
4.4. Describe the operation and construction of counters and registers.
Identify various types of counter circuit. Sketch time related waveforms, draw truth tables.

08, Science background

Sections 001, 003, 004, 006 Number systems, differentiating and integrating circuits, transducers, integrated circuits, transformers and displays.

05, Microprocessor-based systems (Chapter 10)

5.1. Describe the basic principles of microprocessors and microprocessor-based systems. Explain the terminology of the system components, both hardware and software.

5.2. Describe the basic operation and function of associated elements.

5.3. Describe the instruments and techniques used to diagnose and rectify faults in microprocessor based systems. Describe basic assembler instructions.

08, Science background

Sections 001, 003, 004, 006   Number systems, measuring instruments, transducers, integrated circuits, transformers and displays.

08, Science background

(In general, the Science-based material of this syllabus has been integrated with the appropriate technology)

001. Number systems (Chapter 10)

002. Differentiating and integrating circuits (Chapter 6)

003. Measuring instruments (Chapter 1)

004. Transducers (Chapter 8)

005. Electrical principles

006. Integrated circuits (Chapter 3), transformers (Chapter 8) displays (Chapter 9).

Index